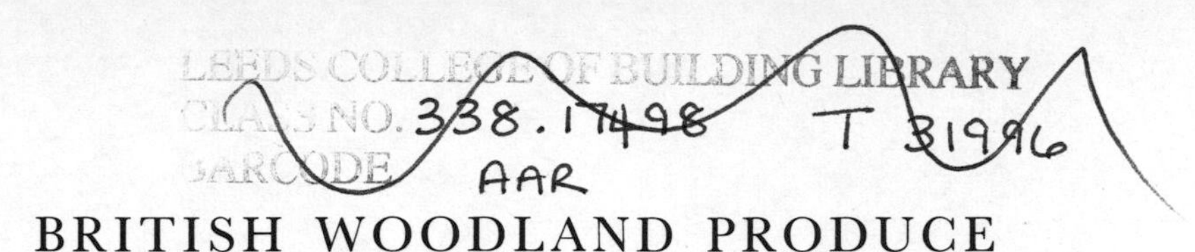

BRITISH WOODLAND PRODUCE

BRITISH WOODLAND PRODUCE

by

J.R. Aaron and E.G. Richards

Illustrated by

Peter Reddy and Elsa Wood

STOBART DAVIES

Published 1990

British Library Cataloguing in Publication Data

Aaron, J. R.
British woodland produce.
1. Great Britain. Forestry industries
I. Title II. Richards, E. G. (Ernest Glenesk), *1918–*
338.17490941

ISBN 0–85442–047–9

Stobart Davies Ltd., 67–73 Worship Street, London, EC2A 2EL

Typeset by Ann Buchan (Typesetters), Middlesex
in 11/12 pt Baskerville
Printed and bound by BPCC Wheatons Ltd, Exeter

CONTENTS

	Acknowledgements	6
	Foreword by Dr F. C. Hummel	7
Chapter 1	Introduction	11
Chapter 2	Uses for Unsawn Wood	28
Chapter 3	Sawlogs and their Conversion to Sawn Timber	43
Chapter 4	The Drying of Wood	58
Chapter 5	Defects and the Grading of Sawn Wood	68
Chapter 6	Fencing	81
Chapter 7	Markets for Sawn Wood other than Fencing	92
Chapter 8	Pests and Preservation	106
Chapter 9	Fuelwood, Charcoal, Wood Gas and Fire-resistance	127
Chapter 10	Properties of the Coniferous Species (Softwoods)	140
Chapter 11	Properties of the Broadleaf Species (Hardwoods)	156
Chapter 12	Residues and Produce other than Timber	167
Chapter 13	The Chemistry and Structure of Wood, and their Implications for Pulp Production	177
Chapter 14	Wood-based Panel Products	193
Chapter 15	Organisations Involved in the Marketing and Use of Wood	202
	Appendix 1 Classification and Presentation of Softwood Sawlogs	209
	Appendix 2 Key British Standards of Relevance to British grown Wood	214
	Bibliography	216
	Index	220

Acknowledgements

The authors wish to thank Mr R. E. Crowther, former principal silviculturist in the Forestry Commission's Research Division, who read and made many useful comments on the draft manuscript; the late Dr K. Wilson formerly Professor at Englefield College, London University who offered valuable observations on Chapters 3 and 13; Mr J. G. Savory formerly principal scientific officer at the Princes Risborough Laboratory of the Building Research Establishment and lecturer at Imperial College, London for help with Chapter 8; and to Mr D. W. Kelley of Carbotech Ltd. for information on charcoal production.

Thanks are also due to Mr G. W. Nunn and Mr A. W. Denby of the Rural Development Commission who gave guidance on the use of timber by craftsmen in Britain.

The authors also wish to thank the following firms, organisations and individuals for providing photographs and illustrations:

Building Research Station
Forestor Ltd
Forestry Commission
Ronald D. Gordon Esq.
G. R. Smith (Engineers) Ltd.
Stenner of Tiverton Ltd
Squirrel Design and Construction
Torvale Products Ltd.

FOREWORD

by Dr F. C. HUMMEL

Head of the Forestry Division of the Commission of the European Communities Brussells (1973–1980)

I welcome the invitation by my two former colleagues Jack Aaron and E. G. (Dick) Richards to write a short foreword to this book because it fills a gap in forestry literature, a gap which they were particularly well qualified to tackle.

A knowledge of product quality and of markets is vital to the success of any enterprise. Forestry is no exception. Most wood goes to sawmills, pulpmills and the panelboard industries, but there are also other existing or potential markets. Examples are high quality logs for veneers, bark for horticulture, transmission poles, stakes, small poles for rustic work and even Christmas trees and ornamental foliage.

This book gives a general account of British woodland produce in plain language; it thus complements the more detailed specialist literature on particular aspects of the subject. In producing the book the authors had in mind the needs of growers, of those engaged in handling, marketing and processing the produce in addition to students of forestry, estate management and wood science. The book is also a convenient work of reference and deserves therefore a place on the shelves of local libraries.

The contents include a description of the wood properties of the broadleaves and conifers commonly grown in Britain and the uses to which they are put. Reference is also made to the micro-structure of wood insofar as it is relevant to the chemical and mechanical utilisation of our forest crops.

The main methods of processing are outlined, namely sawmilling, pulping, building board manufacture and carbonising — together with the preferred species and specifications. The characteristics of the major products are also given, reference being made, where appropriate, to the relevant British Standards. Economically significant insects, fungi and bacteria that attack wood are described together with the necessary preventative and remedial measures. Finally, there is an appendix, a bibliography, a useful list of addresses and an index.

Wood is one of the oldest materials used by man, and in spite of intense competition from other traditional products like iron and steel, as well as from recently developed commodities such as plastics, wood and its derivatives have continued to hold their own. The range and versatility of products made from wood is great. This is due in no small measure to many ways in which it is possible to combine wood with new or improved materials, such as plastics, in the manufacture of paper products and sheet materials like plywood, particle board and fibre building boards. Indeed, the demand for both timber and its derived products in the advanced industrial societies over the next few decades is well assured. Furthermore, new technological advances in the field of wood chemistry are likely to result in the appearance of new products for the future.

The United Kingdom imports some six billion pounds worth of timber and other wood products annually; this represents more than fifty million cubic metres of wood (or a little less than 50 million tonnes). The balance of about six million cubic metres of current consumption comes from woodlands in Britain where production is forecast to rise by fifty per cent by the turn of the century. In the 1950s home production had been a mere trickle.

In the years immediately following the Second World War the plantations that had been created by the Forestry Commission and private growers since 1919 were reaching the stage at which they could be thinned and markets had to be found for the rapidly rising volume of produce from these thinnings. The task was not easy because much of the planting had been of exotic conifers and it was thought that their timber qualities might not match those of trees grown in their native habitats. Laboratory tests soon provided very reassuring results, but potential users remained hesitant. The coal mines in particular, which at that time still consumed large quantities of wood of small dimensions as pitprops, were reluctant to switch from imports to the British product until its reliability had been proved at the coal face. After all, miners' lives were at stake. Forest industrialists for their part initially lacked the confidence to embark on major capital investments in sawmills and other plants to process the unfamiliar produce. The distrust of the potential users of the final products presented another hurdle to surmount.

The traditional broadleaved species presented a somewhat different problem. Wartime fellings had creamed off much of the better timber and markets had to be found for what was left. Moreover, uses had to be found for the produce from the considerable areas of coppice woodland, which was no longer in great demand by rural craftsmen.

The development of forest industries and markets in step with rising production required careful planning. The necessary estimates on future production were prepared by the Forestry Commission while the Forest Products Research Laboratory investigated the characteristics of the individual species. The advice of national and international consultants was sought on general strategy and on the choice of particular technologies. Potential investors were able to use this information for their own feasibility studies. The broad thrust of this work was guided by the Forestry Commission's Home Grown Timber Advisory Committee. The authors of this book were deeply involved in these developments and E.G. Richards also in parallel developments elsewhere in Europe.

Progress has been punctuated by occasional failures and set backs which were caused in part by inexperience and in part by external factors such as the repercussions of the oil crisis of the 1970s and serious fluctuations in the exchange rate which affected the price of imports. But the lessons from these problems have been learned and the 1980s have seen a drastic modernisation and expansion of Britain's forest industries involving massive new investments. Thanks to these developments British sawmills, pulpmills and board industries now rank among the most efficient and profitable in the European Community.

The main motive for further progress in the years to come is likely to be the single European market. Tariff barriers within the EC and between the EC and EFTA have already fallen, but the barriers caused by divergent national product standards will take longer to dismantle.

Small and medium sized firms specializing in particular products are generally expected to benefit most from the expansion of markets through deregulation. Competition will be sharpened by a considerable influx into the EC of Scandinavian forest industrial enterprises. Assurances have been given that the single European market will not be accompanied by a 'Fortress Europe' attitude to outsiders, but many foreign firms in forest industry as in other sectors of the economy believe that the best place to take advantage of the single market is from the inside. Undoubtedly, the influx will have an impact. The prospect for British forestry and forest industry is therefore more severe competition combined with greater opportunities for those who are enterprising, efficient and thoroughly conversant with the potential of British woodland produce.

1

Introduction

The bountiful forest

Forests cover about one third of the world's land surface. In addition to their important function of providing a regular supply of timber and other products, they play other vital roles which include the prevention or reduction of soil erosion, influencing the climate and reducing the "greenhouse effect" by converting carbon dioxide in the atmosphere into wood tissue. In fact, trees represent the most efficient land-based biological system for "locking-up" carbon. At the same time they store the sun's energy and release considerable quantities of oxygen into the atmosphere.

Trees do not require large inputs of fertiliser or protective chemicals (such as insecticides), as do most farm crops, and they can flourish on soils unsuitable for agriculture. Moreover, a substantial proportion of the nutrients which they acquire from the soil is concentrated in their leaves and needles, and is returned to the soil when they are felled for timber. Thus, while some depletion of mineral nutrients may occur when land is used for the production of wood, it takes place at a much slower rate than with other uses.

In some countries the products of the forest other than wood can have even higher values than the timber harvest. Such products include berries and edible fungi, nuts, honey, meat, fodder for domestic animals, furs and skins, tannins, resins, shellac, ornamental foliage and cork. The protection, amenity, recreational and sporting values (such as the sporting rental for pheasant or deer shooting), may in some situations exceed the timber producing values. Yet although the non-wood benefits of the forest are becoming more widely appreciated than hitherto, the primary function of most forests remains the production of wood.

The international dimension

The Second World War caused immense destruction in Europe and the Soviet Union to dwellings, industrial buildings and communications. Although some forests had suffered damage from the actual fighting and others had been over-exploited for the war effort, there was still enough timber to meet the essential requirements of the immediate post-war reconstruction programmes of these areas. The need to secure longer term supplies to meet the continuing and growing needs of the increasing population were recognised internationally and this led to the formation by the United Nations of the Economic Commission for Europe and to the European Forestry Commission of the Food and Agriculture Organisation. Other bodies such as the Organisation for European Co-operation and Development had committees which covered timber and timber products. Through these bodies and their subsidiary technical organisations a great deal of information was exchanged in the fields of forestry *per se*, the forest products industries and the trade in wood products.

The main thrust of forest policies in Europe, the Soviet Union and North America* was directed towards the restocking of areas felled or damaged during the war together with an increase in the total forest area by afforestation of the poorer types of agricultural land. In time the young stands required thinning and consequently there arose a need to find outlets for relatively large quantities of small-sized roundwood, including coppice, over large areas of Europe. This situation, coupled with the realisation that better uses needed to be found for sawmill residues, stimulated the development of new wood-consuming industries such as particle board (wood chipboard) and fibre building board mills, and assisted in the provision of raw material for the pulp and paper industry. The wood processing industries were also helped in their quest for raw material by the decline in the huge demand for pit props in the mines (principally the coal mines) of Europe which followed the mechanisation of mining from the 1960s onwards (see Chapter 2 for British sawn and round pitwood consumption data).

The production of forest products by the major groups within the ECE region (Europe, USSR, and North America) illustrates this trend (Fig 1.1).

In many countries non-native fast-growing species were planted in an attempt to speed-up the production of much needed wood and to improve the economics of forestry. Sometimes the change was from broadleaves to conifers as in the replacement of coppice by conifer high forest; sometimes by the use of species such as poplar for specific purposes such as

* Canada and the United States are members of the Economic Commission for Europe and of the Food and Agriculture Organisation.

ECE Production of Forest Products

Millions of Units (to the nearest Million)

	1950	*1985*	*Percentage Change*[1]
Sawn Wood (m^3)	212	330	+ 56
Wood-Based Panels* (tonnes)	11	81	+633
Wood Pulp (tonnes)	34	113	+234
Paper and Paperboard (tonnes)	40	141	+252

[1] Based on actual, not rounded, figures.

* Wood-based panels include particle board, fibre building boards, plywood, oriented strand board and medium density fibre board (See Chapter 14).

Source "Forest Industries Past and Future" [UN 1987].

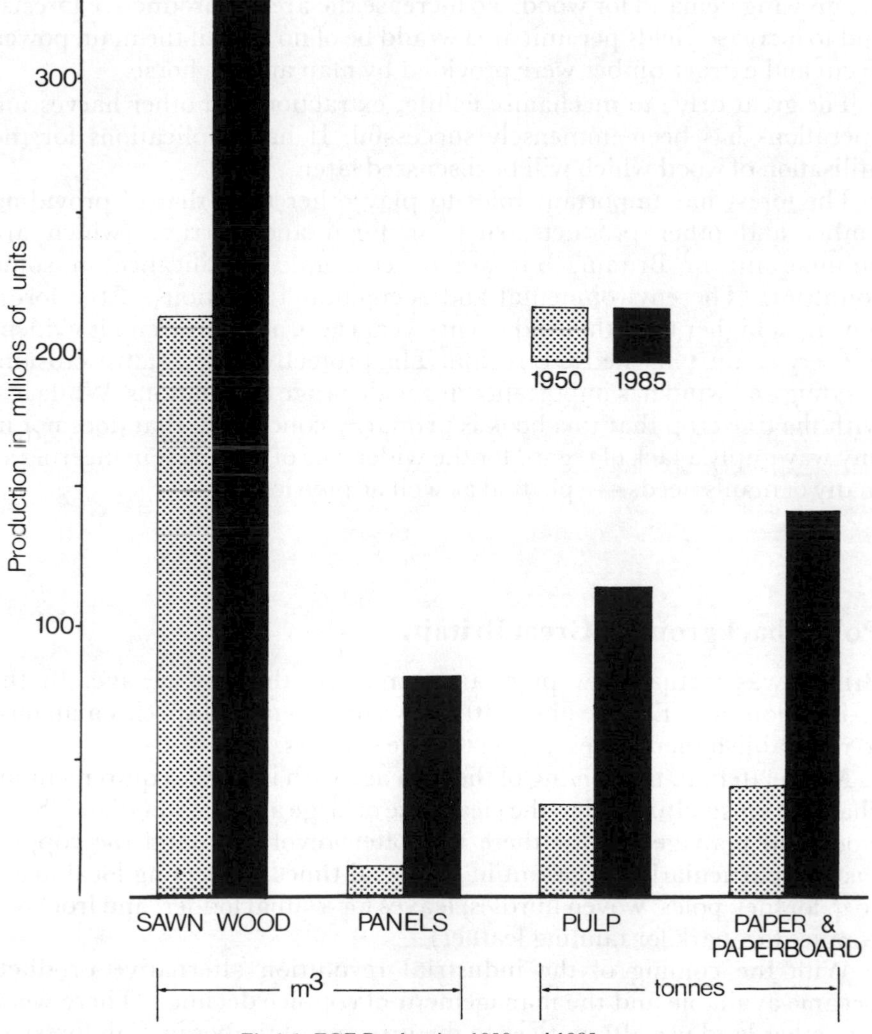

Fig 1.1 ECE Production 1950 and 1985.
Sawnwood, panel products (panels), pulp,
paper and paperboard.

match-making or for pulp production. These policies had implications for forest products research programmes as well as for industry, and, in due course, led to objections from environmentalists who deprecated those changes which altered the ecology of the forests. Some countries (such as Norway and Sweden) have reverted to the use of native species, others have modified their practices and moved away from monocultures, especially of non-native species, in favour of at least a proportion of forests of mixed species.

It was recognised early by both the FAO and the ECE that unless harvesting of wood was mechanised the forests would be unable to meet the growing demand for wood. To increase the area of productive forests, and to increase yields per unit area would be of no avail if the main power to cut and extract timber were provided by man and his horse.

The great drive to mechanise felling, extraction and other harvesting operations has been immensely successful. It has implications for the utilisation of wood which will be discussed later.

The forest has important roles to play other than that of providing timber and other products, such as fungi and berries, (which are unimportant in Britain, but are of economic significance in some countries). The environmental and recreational functions of the forest now rank higher than they did twenty years ago, and this trend is evident in every country in the ECE region. The protective role against erosion, flooding and wind has importance in a wide range of situations. While it is with the tree crop that this book is primarily concerned, that does not in any way imply a lack of regard for the wider role of the forest in meeting so many of man's needs — spiritual as well as physical.

Policy background: Great Britain

Britain was virtually swept clear of trees by the last ice age. In the re-invasion from Europe about 10,000 years ago not all species managed to re-establish themselves, especially the conifers.

Much later on, the coming of the iron age with its high requirement for charcoal for smelting led to the clearance of large areas of woodland. Such woodland management as there was often revolved round the coppice system, particularly important in medieval times as meeting local needs (e.g. for fuel; poles; woven hurdles; leaves for animal fodder; and from oak coppice tan-bark for tanning leather).

With the coming of the industrial revolution alternative products became available and the management of coppice declined. There were, too, other land use alternatives to the growing of timber in high forest — such as sheep in upland Britain, especially in the Scottish Highlands.

These and other factors combined to make Britain one of the most poorly wooded of any temperate country. There was little economic pressure for change. Throughout the nineteenth century British woodlands had to compete with the low-priced high quality coniferous timber that was readily available from Scandinavia, Russia and Canada, in addition to the high quality hardwoods which were arriving from a trade conscious empire. There was little or no incentive to plant or manage woodlands for timber production. Attitudes began to change during the First World War when a most effective enemy blockade caused an abrupt realisation that Britain was almost wholly dependent on imported timber supplies, to the extent that shortages of essential wood might well have brought defeat on the "production front" rather than on the battlefield.

The total woodland area at the outbreak of that War was 1.29 million hectares (5.7 per cent of the land surface), most of which was productive. By 1924 it had been reduced to 1.25 million hectares of which only two-thirds was productive, most of the felling having taken place during the war years.*

The necessity for a strategic reserve of well-managed woodland, which would ensure sustained supplies in the event of another national emergency, had become all too apparent, with the result that the Forestry Commission was set up in the wake of the Acland Report (1918) on post-war reconstruction** to establish new plantations and to encourage planting by private estates.

The advent of the Second World War caused further serious depletion of the woodlands and, in consequence, most of the coniferous high forest remaining after this war was below middle age. Hence the Forestry Commission was required to increase rapidly its rate of acquisition of land for planting — mainly new planting — and to introduce new incentives for private owners to replant their felled woodlands and to afforest new areas. The national strategic reserve was quantified as an overall target of two million hectares of productive woodland.

In 1958 the Government announced that, as part of their general review of defence policy, a strategic reserve of standing timber was no longer a *prime* defence consideration. Although the target of two million hectares was not formally abandoned, the Forestry Commission's planting programme was to be set for a period of ten years ahead and then reviewed. Several reviews have indeed taken place, at varying intervals of time, but in the event planting and restocking by the Forestry Commission and by private owners, which amounted to some 200,000 hectares in the period 1919–39, totalled 592,000 hectares in the period

* Forestry Commission Census Report No 1 (HMSO) 1952

** Final Report of the Reconstruction Committee — Forestry Sub-Committee, Ministry of Reconstruction 1918 (Cmd 8881).

1946–65, and reached 706,000 hectares in the period 1965–84; in all a substantial achievement.

The total woodland area is now 0.6 million hectares of broadleaved forest (including a small area of productive coppice) and 1.5 million hectares of conifers. This not inconsiderable planting programme has been achieved under policy objectives which have varied somewhat with the passage of time as different governments have laid stress on this or that particular aim. At one time the balance of payments argument (import saving) was considered important. The provision of employment, especially in rural areas where few other forms of land use offered a viable alternative to forestry, has been a common thread running through all the various Government policies. The economic argument — the need for forests to provide the (mainly coniferous) wood for industry in a cost effective way — has also been a common feature of forest policy reviews. In adopting these aims Britain was very much in line with its Continental neighbours (eg France). The most obvious result of pursuing these policies has been to favour the planting of exotic conifers such as Sitka spruce, Corsican pine and Douglas fir, which have faster rates of growth and better log form than the native Scots pine, and which can produce worthwhile crops on comparatively infertile soils so often available to forestry.

In the last decade there has been a strong movement, in Europe as well as in Britain, to give more emphasis to the environmental aspects of forestry. The adjustment of fiscal policies to give incentives — hitherto lacking — to the planting of broadleaves has been a case in point.

The plantations

The greater part of the plantations created as a result of the forest policies of the past 40 years, consist of one or more of nine major species of conifer that grow well in Britain, namely Scots pine, Corsican pine, lodgepole pine, European larch, Japanese larch, Dunkeld larch, Sitka spruce, European spruce and Douglas fir. Large areas are monocultures, especially where Sitka spruce has been planted as the optimum volume producer within the desired rotation. Mixtures of two species of conifer were not uncommon in the early days of the Forestry Commission (eg a pine and a spruce, or Japanese larch with Sitka spruce). Conifer broadleaf mixtures have enjoyed a renewal of interest. While a mixture of species adds to the complexity of plantation management and marketing the produce, there is evidence that it can help to maintain the long-term fertility of the site as well as improving the aesthetic appeal of the forest and hosting a more varied wild-life.

The choice of broadleaves for planting has favoured much more the use of native species such as ash, beech and oak. Sycamore, strictly not a native, has been widely planted, especially in northern England and parts of Scotland. On selected sites poplars have been grown for specialised uses; and there is now increased interest in growing the native wild cherry and, in Scotland, birch.

Although there are a number of options open to the forest manager which will ultimately affect both the shape and diameter of the round timber at the time of harvesting, as well as the properties of the sawn wood, more often than not only two such decisions are made, namely the initial spacing at the time of planting, and the subsequent frequency and intensity of thinning. If the distance between plants is increased significantly the quality of the final crop is reduced. On the other hand close-spacing, which requires a greater number of plants per hectare, as well as a higher labour input, is usually more expensive. The timing and intensity of thinning influences the characteristics of the wood grown in the stand. Closely planted stands of softwoods (eg with a spacing between one and one and a half metres), if left unthinned will produce logs of smaller diameter, less taper, smaller diameter knots and a narrower core of juvenile wood, than stands at two metre spacing which have been thinned regularly. Regular thinning also means that the sawlog sized timber is produced earlier in the rotation than in unthinned or lightly thinned stands. Thus, although thinning intensity has virtually no influence on the total production of wood (provided that it is not excessively heavy), it does influence how the wood arises — a few large stems or many small diameter stems.

The table below shows two cross-sections of Sitka spruce cut from sawlogs (at a height up the tree equivalent to five years in age). Section A comes from a stand of trees close-planted, which suffered a check in growth until the canopy closed at an age of 17 causing the competing heather to be suppressed. The trees then grew rapidly, and subsequent frequent and heavy thinning maintained the diametric growth at four to five rings per 25 mm of radius. Tree B was planted at wider spacing than tree A and made good early growth, but no thinning was undertaken with the result that the rate of diameter growth fell to 12 rings per 25 mm from the age 16 to 28 years.

	mm from the pith:			*25*		*50*		*75*		*100*
Tree A	No of rings:		12		4		4		5	
	Age in years:	5		17		21		25		30
Tree B	No of rings:		4		3		4		12	
	Age in years:	5		9		12		16		28

Most methods of thinning aim at removing badly formed trees (ie crooked stems, forked stems, coarsely branched trees) to improve the overall quality of the stand by concentrating the growth potential of the site on the best trees. Removal of trees of inferior form can also influence future stands by removing as potential parents those trees whose defects are inherent; this aspect has become less important than it once was because seed from which new plants are grown can now be obtained from seed orchards where a high degree of control is exercised over the parentage of future crops.

The cost of carefully selecting and marking the trees to be removed in softwood stands is high, and the value of the yields from the early thinnings is relatively low. A substantial reduction in the cost of early thinning can be achieved by removing whole lines of trees (line or mechanical thinning) irrespective of their size and quality. It has the additional advantage that it facilitates access by harvesting machines, particularly where two adjacent rows are felled. For a variety of reasons it is usually cost-effective to have subsequent thinnings individually selected and marked by a skilled forester.

The same basic principles apply to the thinning of broadleaved crops. Their slower rates of growth means that the first thinning will normally be carried out at a later age than in conifers. Many woodland owners are reluctant to depart from the traditional practice of the individual selection of the trees to be removed in the first thinnings in hardwood stands.

Windthrow is a major problem on an island which is frequently subjected to gale-force winds, and where shallow-rooting tree crops are often planted on soft peaty soils with poor cohesion. This hazard can be reduced considerably if the plantations remain unthinned, thus ensuring that the mutual support of the trees is optimised. The incidence of windthrow can also be lessened by growing the crop on short rotations so that the trees are felled before they reach heights at which they are susceptible to wind. Many forests in upland areas have been subjected to non-thin regimes during the past two decades, consequently a significant proportion of the coniferous wood due to be marketed in the early part of the 21st century will be from crops which have never been thinned.

Prior to 1973 the grant aid system for private planting offered no special encouragement for the growing of broadleaved tree crops (90 per cent of the total area of broadleaved woodland is in the private sector). In 1973 broadleaf planting and maintenance grants were increased relative to conifer grants. After wide consultation, the grant aid policy was further amended in June 1985 with the objective of maintaining and enhancing the value and character of Britain's broadleaf woodlands, as well as placing greater emphasis than hitherto on the retention of "ancient and semi-natural broadleaf woodland". The implementation of this policy is by increasing planting grants for the establishment and replenishment of broadleaf crops, together with stricter controls on broadleaf felling. The felling control policy will bring a reduction in the volume of home-grown hardwoods coming on to the market for at least the next four decades.

The wood resource

There are currently some 2.1 million hectares of productive woodland in Britain of which 0.9 million hectares are held by the Forestry Commission and 1.2 million hectares are privately owned. Approximately 1.5 million hectares are coniferous and 0.6 million hectares broadleaved high forest. There is also a small area of productive coppice — mainly sweet chestnut, introduced by the Romans, and the native hazel — almost all of which is in private ownership.

Annual removals of *coniferous* timber are estimated at:

	1985	*1990*	*2000*	
Forestry Commission	2.9	3.4	5.4	Millions of cubic metres
Private Woodlands	1.2	1.7	3.4	(overbark measure)
TOTAL	4.2	5.1	8.8	

N.B. The above totals may not add up due to rounding.

Broadleaf removals, mostly from private woodlands, are expected to remain constant at around 0.9 million cubic metres per year, which is slightly below the 1980–85 average.

Recorded deliveries to industry totalled some 4.5 million cubic metres (underbark) in 1985, of which softwood sawlogs were 2.4 million cubic metres, hardwood sawlogs 0.5 million cubic metres; chipboard and fibreboard material 0.4 million cubic metres; pulpwood for use in Britain 0.6 million cubic metres (with 0.35 for export); and other uses 0.15 million cubic metres. In addition, the recorded deliveries of fuelwood were of the order of 0.13 million cubic metres.

Harvesting

In this section brief mention is made of the harvesting systems in current use because the choice of system influences and in turn is influenced by the markets being served (eg poles; sawlogs; pulpwood; board mill material; pit props; or fencing timber). The type of terrain over which the extraction takes place using cross-country vehicles (eg tractors) or using cableways, and the density of the forest road and track network, will also influence the choice of harvesting system. In Britain the load carrying capacity of minor public roads may also be important; in some regions these may have been built to a standard lower than that to which forest roads are commonly built. The interaction between these and other factors can be complex, and what follows here can only summarise in a somewhat simplistic way the relationship between harvesting systems and the uses to which the timber from the thinnings or final fellings are put.

A crucial question which has to be faced is the amount of time which can be devoted in deciding how to cross-cut the felled tree to ensure the optimum revenue from its sale. In the more valuable hardwoods the price differential between a veneer butt, planking quality, fencing quality and mining timber may be of the order of 20:5:2:1 for oak in southern England. The dividing lines between these types of roundwood is often sharp and readily recognisable to the expert eye, and the large price differences in the products fully justifies time being spent on deciding where to cross-cut.

Some conifers which have special uses such as boatskin larch can show comparable differences in price between adjacent parts of the stem, but by far the greater bulk show a *gradually reducing* value per cubic metre as the diameter decreases from butt to tip. The exact point at which the first sawlog is cut from the butt of a mature stem is usually important, but beyond this point the decision as to where to cross-cut becomes progressively less significant, and a trained forest worker should be capable of dividing up the tree to optimise its value according to the prevalent market values.

Harvesting costs rise as the number of assortments cut from a crop increases. This is true even in forest depots with fully mechanised cross-cutting, but in a depot the extra cost of producing, say, four assortments as compared with two is not great, and increased revenue can be obtained by maximising the out-put of higher value assortments.

Over the past thirty years the harvesting of timber has been almost completely mechanised. In the most advanced systems — suitable only for very large-scale operations — the operators of sophisticated machines, which fell, delimb and cross-cut, remain in air-conditioned cabs and can continue to work during unfavourable weather. They carry their own flood-lighting for night shift work. At the other end of the scale, felling, delimbing (snedding) and cross-cutting is by power saws, followed by extraction by tractor or forwarder. This method is suited to the requirements of small scale harvesting. A recent project* has re-examined the possibility of using horses for extraction of timber, particularly where terrain is difficult and woods small. Although horses still play a part in wood harvesting operations in some countries (eg Sweden), it seems unlikely that their use in more than a handful of instances will occur in Britain.

There are three basic harvesting systems to which many variations can be applied. They are (Fig 1.2):

– the shortwood system
– the tree length system
– the full tree system

* Coed Cymru — Small Wood Project — Countryside Commission

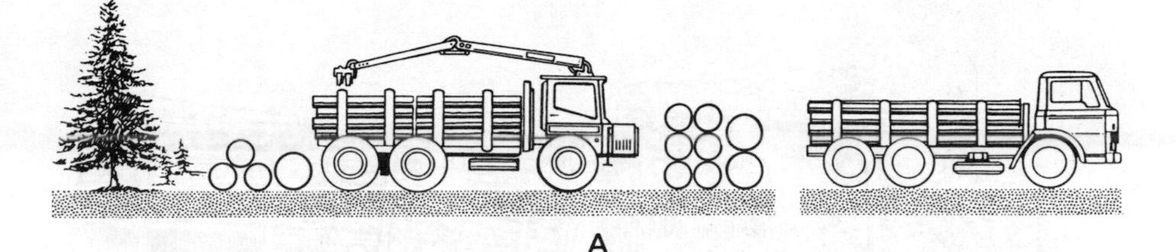

A

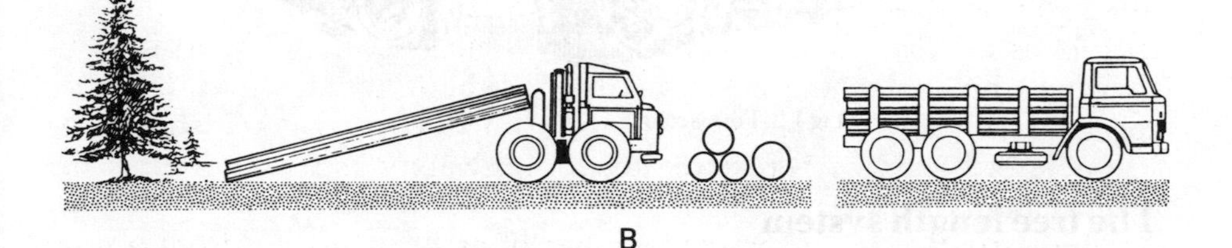

B

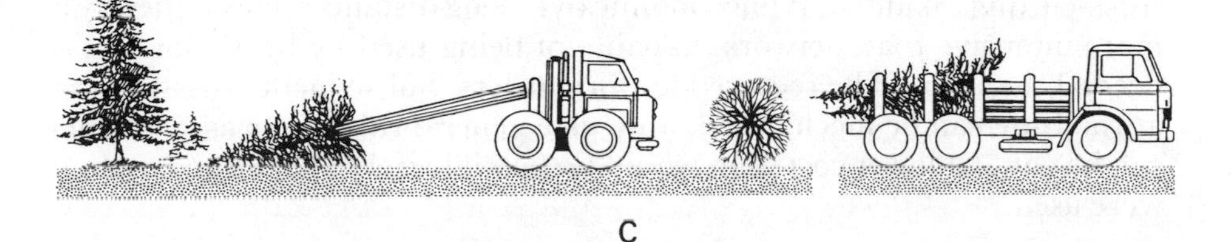

C

Fig 1.2 *A*, Short Wood. Convert at stump. Forwarder to roadside.
B, Tree Length. Debranch at stump. Skid to roadside. Convert at roadside.
C, Whole tree. Skid whole tree (with branches) to roadside. Debranch and convert at conversion depot.

The short wood system

The modern short wood system is designed to make full use of forwarders (Fig 1.3). These are relatively fast-moving cross-country vehicles capable of carrying loads of up to 15 tonnes. They are usually equipped with a hydraulic grapple crane for loading and unloading. It is economic to use them over considerable distances, travelling over the forest floor, tracks and unmetalled roads. In this system the felling, snedding and cross-cutting are carried out in the stand either by power saw or processer, after which the converted produce is conveyed by the forwarder to forest roadside for loading on to a main road lorry.

Fig 1.3 Forwarder

The tree length system

The primary objective of this system is to minimise the work undertaken within the stand. The trees are felled and delimbed by power saw or machine before being skidded in their full length to forest roadside for cross-cutting. Skidding is uneconomic over long distances, consequently a more intensive road network, capable of being used by heavy lorries, is needed than would be required for forwarders. Subsequent cross-cutting at roadside can be mechanised, and, as a general rule, more assortments can be cut at a lower cost than would be possible if the short wood system were used.

The full tree system

In this system the whole tree is skidded to the forest roadside or depot for conversion. A high degree of mechanisation is possible and a larger number of assortments can be cut to optimise the yield without incurring a financial penalty. In some situations the lop and top can be utilised together with the off-cuts as fuelwood, thus increasing the cash return. Where no outlets for the lop and top exist large accumulations at depot have to be avoided on account of the high fire hazard. Solutions include bundling such residues and bringing them back to the forest as return loads; or burning in incinerators. In Britain with its relatively dense population and comparatively dense traffic on its public roads, the use of the very long lorries needed to carry whole trees to a conversion depot outside the forest is not a practical proposition. The use of the lop and top as fuelwood is seldom economically feasible under British conditions. Whole tree logging to depots remote from the forest is therefore not currently practiced.

Fig 1.4 Timbermaster Skyline at work with Forestry Commission, North Scotland Conservancy. Photo courtesy G. R. Smith (Engineers).

A further disadvantage of the system which can be significant on the less fertile sites is the long-term impoverishment of the forest soil through the removal of foliage which contains an appreciable proportion of the nutrients which would otherwise be recycled when delimbing is carried out in the forest.

Cable-cranes

Where the terrain is too steep for tractors to operate extraction is by cable-crane. All three of the above systems can be operated using cable-cranes. (Fig 1.4).

Harvesting systems and the value of the converted produce

The choice of harvesting system, as already noted, is influenced by a number of factors which affect the quality and value of the converted produce. These include:

Method of felling Power saws make a clean cut at right angles to the stem, while hydraulic secateurs may cause splits at the butt end or cut it at an angle.

Cleanliness Where skidding is used on delimbed stems and logs the bark will inevitably be contaminated with mud and grit (except on peat soils), causing excessive wear on debarker blades and other machinery. Where conditions permit, it is clearly advantageous to retain the branches until after extraction.

Cross-cutting It is usually more difficult, especially in sloping terrain, to sustain accuracy in cross-cutting to the required lengths at right angles to the stem at stump than at forest roadside or in a depot. Where cross-cutting is undertaken by a processor or similar mobile machine accuracy is not dependent on the location.

Storage Any lengthy delays in conveying the roundwood from forest to mill can increase the risk of damage by staining fungi and ambrosia (pin-hole) beetles, especially during the warmer months. Delays can also lead to unwanted drying of wood which is particularly disadvantageous where it is required for pulping by a mechanical process. On the other hand, prolonged storage in the forest, especially when the bark has not been removed, can lead to drying rates which are too slow for such uses as pit props and wood wool, and can eventually result in decay.

Hygiene

Neglect of stump treatment in the thinning stage can lead to serious loss of timber through butt rot in the main crop. Thus, under British conditions, it has become virtually essential for harvesting operations to be accompanied by the hygienic measure of stump treatment. This is because the parasitic fungus *Heterobasidion annosus*, which in addition to causing butt rot, is capable of attacking and killing young trees, uses the stumps of freshly felled trees as a source of nutrition and subsequently as a base from which it invades the roots of nearby healthy trees. It is therefore standard practice to treat the newly exposed stumps with a chemical such as urea which inhibits fungal growth, or with an innoculate of a competing fungus such as *Peniophora gigantea*.

As a routine measure of protection in pine plantations, unbarked logs should be removed quickly to preclude attack by the pine shoot beetle (*Tomicus piniperda*). In stands of spruce prompt removal of felled timber may be necessary to prevent damage by ambrosia beetles (See

Chapter 8).

Hygiene measures also extend to the transport of timber, and regulations are sometimes imposed to prohibit the movement of unbarked wood. The outbreak of the Dutch elm disease (caused by the fungus *Ceratocystis ulmi*) in the 1970s and the subsequent discovery in the 1980s that the great spruce bark beetle (*Dendroctinus micans*) had attacked spruce in Britain, resulted in the enforcement of restrictions on the movement of unbarked timber which are still in operation in 1990.

Uses, markets and marketing

There is a clear distinction between the possible uses for which a species of timber may be put and the existance of a market. The objective in this book is to concentrate on current uses for woodland produce which are of some economic significance, but in a period of rapid technological change, some uses will disappear under the impact of competition from alternative materials and products, while other uses will be developed. Progress in manufacturing techniques can also mean that specifications change with regard to both the acceptability of species and the sizes that are suitable.

Experience over many years has shown that there are some species of both conifers and broadleaves which are in continuous demand by a wide range of customers for a wide range of uses. These include in alphabetical order:

Softwoods (Conifers)

Douglas fir *Pseudotsuga menziesii* (Mirbel) Franco

Larches
- European *Laris decidua* Miller
- Dunkeld [hybrid] *Larix x eurolepis* Henry
- Japanese *Larix kaempferi* Lambert

Pines
- Corsican *Pinus nigra* var. *maritima* Aiton
- Lodgepole *Pinus contorta* Douglas
- Scots *Pinus sylvestris* L.

Spruces
- European *Picea abies* (L) Karsten
- Sitka *Picea sitchensis* (Bongard) Carriere

Hardwoods (Broadleaves)

Ash *Fraxinus excelsior* L.

Beech *Fagus sylvatica* L.

Elm *Ulmus spp.*

Oak *Quercus robor* L. and *Q. petraea* Lieb.

Poplar *Populus spp.*
Sweet chestnut* *Castanea sativa* Miller
Sycamore *Acer pseudoplatanus* L.

* especially as coppice.

Other species referred to in this book, which are unlikely to be commonly and regularly available, include:

Softwoods (Conifers)

Lawson cypress *Chamaecyparis lawsoniana* (Murray) Palatore

Leyland cypress *Cupressocyparis leylandii* (Jackson and Dallimore) Dallimore

Pines
- Austrian *Pinus nigra* var. *nigra* Harrison
- Maritime *Pinus pinaster* Aiton
- Radiata *Pinus radiata* D. Don
- Yellow *Pinus strobus* L.

Sequoia *Sequoia sempervirens* (D. Don) Endlicher

Silver firs
- European *Abies alba* Miller
- Grand *Abies grandis* L.
- Noble *Abies procera* Rheder

Wellingtonia *Sequoiadendron giganteum* (Lindley) Buchholz

Western hemlock *Tsuga heterophylla* (Rafinesque) Sargent

Western red cedar *Thuja plicata* D. Don.

Yew *Taxus baccata* L.

Hardwoods (Broadleaves)

Alder *Alnus* spp.*

Birch *Betula pendula* Roth and *B. pubescens* Ehrh.*

Eucalyptus *Eucalyptus spp.*

European cherry *Prunus avium* L.

Holly *Ilex aquifolium* L.

Hornbeam *Carpinus betulus* L.

Horse chestnut *Aesculus hippocastanum* L.

Laburnum *Laburnum anagyroides* Medic

Lime *Tilia spp.*

Plane, European *Platanus acerifolia* Willd

Robinia *Robinia pseudoacacia* L.

Southern beeches
- Coigue *Nothofagus obliqua* (Mirbel) Blume
- Rauli *Nothofagus procera* (Poepig and Endlicher) Oersted

Walnut *Juglans regia* L.

Willow *Salix spp.*

* These species may be available in quantity on a regular basis in turnery pole sizes.

Even where a market exists for a particular type of wood, it by no means follows that it is open to all potential suppliers. Some firms will give preference to those producers who can offer continuity of supply, and who have proven reliability in meeting delivery deadlines and customer specifications. Before undertaking harvesting of timber, it is prudent to ascertain what markets exist for the species, sizes and qualities that the woodland will yield and secure firm orders.

Variability in wood

In common with other natural products, wood exhibits considerable variability in all of its properties. Generations of craftsmen have chosen their material on the basis of skill and experience, and structural engineers and statisticians have refined the selection process. While this variability adds to both the interest, and, indeed, the aesthetic appeal of timber compared with other building materials, it does, of course, necessitate the use of special sampling techniques for the selection of the material for the testing for strength and other properties. It also requires the application of appropriate statistical methods to the test data to ensure that the maximum amount of reliable information is obtained from them, and that the results are evaluated and presented to potential users in a meaningful way.

Although the starting point for an assessment is obviously the arithmetic mean, the variability around the mean is usually so great that this value on its own has limited application. For example, it would be unwise and probably untrue to assert that because the mean compressive strength of pine pit props grown in East Anglia is roughly twelve per cent higher than pine pit props grown in the north of Scotland, East Anglia invariably produces stronger pine. It is outside the scope of this book to describe, in a statistically explicit way, the techniques required for sampling populations of the individual species for strength and other property tests. Suffice it to note that the quantification of variability is expressed as either the *variance** of the population, or its square root — the *standard deviation* — and that the quantified results of timber research are of little value unless one or the other is quoted with the mean.

* The variance of the population is calculated by (1) taking the squares of the differences of the individual samples from the mean, (2) adding them together, and (3) dividing by the number of samples minus one.

2

Uses for Unsawn Wood

This chapter describes the most common uses for timber in the round, except for fencing material which is dealt with in Chapter 6. There are few markets for round wood less than 7 centimetres in diameter (over bark), and trees below this size are not normally harvested, although under some conditions it can be economical to chip small stems and branchwood in a mobile chipping machine and to market the product.

There are a small number of premium markets for timber which is to be used in the round; these include telegraph poles and poles for electricity supply (Fig 2.1). There is a somewhat smaller market for structural round timber and for flag poles, rugby football goal posts and civil engineering works. On the other hand the fencing and pitwood markets use a large quantity of small diameter round material.

Heartwood and sapwood

Although not normally included in log grading systems the relative proportion of heartwood and sapwood may affect the suitability of logs for various uses, and this may be reflected in the prices obtained. A low proportion of the paler sapwood would be desirable in the selection of, for example, oak veneer logs, while a wide band of sapwood would be favoured in Scots pine telegraph poles, because it increases retention of preservative and the depth of radial penetration.

Sapwood fulfils the triple role of conduction of water and mineral nutrients from the roots to the foliage, contributing substantially to the mechanical support of the crown, and the storage of reserve food materials. Its tissue contains living cells and dead cells, which are fully saturated and permeable to the passage of fluids.

Heartwood consists entirely of dead tissue, and the only purpose which

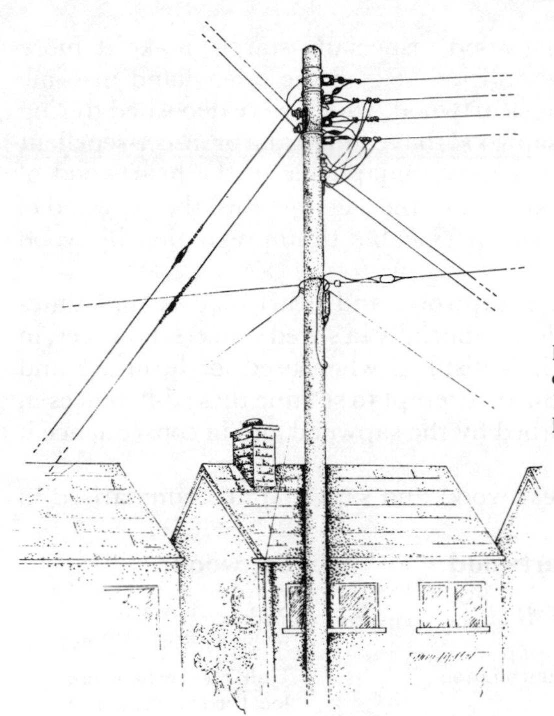

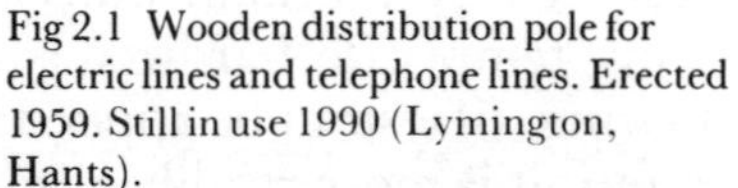

Fig 2.1 Wooden distribution pole for electric lines and telephone lines. Erected 1959. Still in use 1990 (Lymington, Hants).

it serves is that of mechanical support. Its cells are often relatively impermeable to the passage of fluids. The moisture content, though high, is in most species well below that of the sapwood. It is frequently, though not necessarily, darker than the sapwood (in oak and Scots pine the distinction is quite clear, but in spruce and birch it is less noticeable). There is no pre-determined age at which the sapwood of a species changes into heartwood, and the boundary between them is often irregular, occurring at different ages within the stem; this means that it frequently crosses the annual rings without any obvious cause.

While the radial widths of sapwood tend to be characteristic of a species (for example, the band of sapwood is normally much wider in oak than in sweet chestnut, and wider in European spruce than in Sitka spruce) the extent of the sapwood is usually a reflection of the vigour of the individual tree (see page 33).

In the growing stem decay sometimes occurs in the heartwood, where it is manifest as butt rot or pipe rot; this is possible because there is sufficient air available to favour the development of wood-destroying fungi. Such decay rarely occurs in the sapwood because there is too much moisture and too little air — although harvesting damage to the stem can permit localised drying to occur and the entry of decay fungi (see Chapter 8). The situation is reversed after the timber is converted and dried, or partially dried.

The nutrients present in sapwood, especially starch, make it more susceptible to attack by fungi and insects; on the other hand in some species substances found in the heartwood, which were deposited during the course of the change from sapwood, have fungicidal or insect-repellent properties. The substances known as thujaplicins in the heartwood of western red cedar afford a striking example. Against this, the sapwood of most species is appreciably more permeable to impregnation by wood preservatives.

The colour contrast between sapwood and heartwood is sometimes exploited for its ornamental effect, especially in sliced veneers; however, in many temperate hardwoods it is disliked when used for furniture and flooring, (if a stain be applied in an attempt to subdue these differences in colour, it is more readily absorbed by the sapwood and in consequence it may appear darker).

The differences between heartwood and sapwood are summarised in the table below:

	Heartwood	**Sapwood**
Function in growing tree	Support	Support, upward conduction and storage
Decay in standing timber	Not uncommon	Unlikely except where localised drying may have occurred in the wake of injury
Decay in converted timber	Many species resistant to varying degrees	Susceptible
Stain in felled timber	Immune from blue (sap) stain, other stains unlikely	Susceptible, especially in pines and Douglas fir
Lyctus beetle attack	Immune	Hardwoods with large pores susceptible
Pine hole borer (Ambrosia beetle) attack	Immune	Some species susceptible
Furniture beetle attack	Resistant, but not immune	Susceptible
Strength of clear (defect free) wood	Equal to sapwood except where a significant proportion of juvenile wood is present (see page 77)	Equal to heartwood
Permeability to fluids including preservatives	Many species resistant	Most species permeable
Colour	Frequently darker, but little contrast in lighter timbers	Often appreciably paler
Tyloses (in the vessels of some hardwoods)	May be present (see page 184)	Absent

Poles

Considered on a world basis there are four main markets for poles:

- Telecommunications*
- Electricity supply*
- Structure, especially for agricultural buildings (pole barns)
- Piling for river and sea defences, jetties and wharves

For each of the above uses, durability (ability to give forty years service without remedial treatment against decay or wood-boring organisms), strength (especially stiffness and resistance to shock loading) and straightness are important. The need for straightness may be to facilitate the fixing of other members to the poles, e.g. sawn horizontals in jetties and sea defence works and pole barns, as well as to meet standards of appearance in the finished structure.

In Britain straightness is considered important from the aesthetic point of view in transmission poles, although many other countries tolerate less straight and therefore cheaper poles.

Telegraph and power transmission poles

Scots pine (and its imported equivalent, redwood) is the traditional species used by both the telecommunications and electricity supply industries, because it has the required strength properties, and because its permeable sapwood gives the desired retention of preservative (usually creosote). Moreover, Scots pine poles meeting the specification for size and straightness are available in large numbers from Scandinavia (especially Finland) from where they can be shipped to those UK ports where yards for their further preparation have been established.

There was a surge in demand for poles during the post-war years, resulting not only from an extensive programme of rural electrification during the late 1940s, but also from a substantial increase in telephone rentals particularly in rural areas. This in turn brought about an intensification of interest in the greater use of home-grown poles to augment supplies and to help to reduce the national trade deficit. This led to close cooperation between the Forestry Commission and the Post Office† during the immediate post-war decades and to joint research between the Forestry Commission, the Electricity Council and certain other organisations during the period 1977–83.

* In the UK single poles often carry lines for both telecommunications and electricity supply [see Fig 2.1].

† Before the setting up of British Telecom and Mercury in the 1980s the Post Office was responsible for public telecommunications [except in Kingston-upon-Hull].

Much of the cooperation between the interested parties took the form of development work to consider the possibility of using home-grown coniferous species other than Scots pine, especially those which are normally faster growing and capable of producing a higher yield of straighter poles per hectare. One result was that the scope of British Standard 1990:1984 "Wood poles for overhead lines (power and telecommunications)" was extended to facilitate the use of the major coniferous species. The advantage and disadvantages of alternative species to Scots pine are summarised below:

Corsican pine has properties comparable to Scots pine, but it has a superior stem form resulting in an appreciably higher yield of acceptable poles per hectare and it has a much greater proportion of sapwood which permits a far higher retention of preservative. The diameters of the knots tend to be greater than in other species, thus special care is needed in the selection of poles to avoid large knot whorls. It is notoriously difficult to dry (it takes longer than other species and it can actually absorb moisture during the late autumn and winter months) so that the possibility of deterioration starting before preservation treatment has been applied has to be borne in mind. This risk can be avoided if the poles are treated by a sap-displacement method (see page 34).

Douglas fir was used for many years for telegraph poles, principally in south western England and south Wales. Its strength properties are marginally better than those of Scots pine and although it has greater natural resistance to decay, it is more difficult to treat with preservatives under pressure. The yield of poles per hectare is usually low.

The larches (European, Japanese and Dunkeld larch) are significantly stronger than Scots pine (partly because the knot whorls are smaller, and partly because the wood substance is inherently stronger). It is also more durable, but resistant to pressure treatment with preservatives. All three species have a low yield of suitable poles per hectare.

European spruce was used for telegraph poles until the 1960s, but it is not currently used for telecommunications. However, following the recent development work, a number of poles have been put into service for electricity supply. It is roughly 15 per cent weaker in resistance to bending than Scots pine; consequently stouter poles are needed to give a comparable performance. Once dried, it is extremely difficult to obtain satisfactory penetration of the sapwood by preservatives applied under pressure. In the undried (green) condition spruce sapwood is no less permeable than that of the pines in that it permits free diffusion of fluids. This means that sap-displacement methods of treating fresh-felled spruce, using either copper-chrome-arsenate or an emulsion of creosote, are quite effective and offer a number of advantages (see page 34). The yield of poles per hectare is moderately good.

Sitka spruce poles have a bending strength of nearly 20 per cent lower than that of Scots pine, with the result that larger diameters are needed for the same performance. As with European spruce the sapwood is permeable only in the fresh-felled condition and once drying has taken place it is extremely resistant to penetration by preservatives. While sap-displacement methods can be employed to give a satisfactory retention of preservative, the proportion of sapwood is often too low to ensure an adequate loading. Careful selection in the forest is therefore necessary to ascertain that the radial depth of sapwood is at least 50 mm (it can be as low as 7 mm which is of course quite inadequate). The identification of the sapwood can be made in a number of ways using chemical dyes such as bromocresol green, or by a "smudge test" with an indelible pencil, but the most practical method for the harvesting operative is to cut a thin (5 mm) disc from the base of those poles which are seen to conform with British Standard 1990 in respect of size and form. When the disc is held to the light, the sapwood is translucent while the heartwood remains opaque. The thickness of the sapwood is a reflection of the amount of green crown on the standing tree. Sylvicultural techniques (such as heavy thinning and fertiliser application) which increase the quantity of sapwood can therefore be adopted. Provided that a minimum 50 mm radius of sapwood is present, the yield of suitable poles per hectare is moderately good.

The preservation of poles

The preservation of both telegraph and electricity supply poles is normally carried out by applying creosote under pressure in cylinders located at the yards of the pole importers (these are at Leven [Fife]; Bo'ness [Lothian]; Blyth [Northumberland], Fleetwood [Lancs]; Boston [Lincs]; Belvedere [Kent]; Newport [Gwent] and Grange Court [Glos.]. The method employed is usually the Rueping empty cell process in which a final vacuum is drawn in the treatment cylinder to remove excess preservative. The exudation of creosote while the poles are in service (known as bleeding) can be troublesome; and it had been observed that it tended to be more severe in poles which had been transported by road than in poles which had been floated down rivers to the port of loading (at one time a common practice in Scandinavia). This phenomenon led to the view that poles which had been floated were more permeable, and that the improvement in permeability was the result of bacterial activity in saturated poles. Subsequently it was demonstrated by the former Forest Products Research Laboratory, that the sapwood of a species such as spruce, which when dry is difficult to penetrate by preservative, can be rendered permeable by immersion in water for several weeks. This work was followed up by intensive research in the Irish Republic which established that the increased permeability was indeed the result of the digestion by bacteria of the pit membranes in the walls of the wood tissue. The work was taken further in a joint project by the Electrcity Supply Industry, the Forestry Commission and others. However, after extensive trials it was concluded that, although water-treatment by either ponding or spraying improved both the penetration and retention of preservative significantly, it was not feasible as a commercial operation, mainly because it prolongs the treatment cycle substantially, and because it does not improve the permeability of the heartwood.

The researchers pursuing the joint project also investigated the possibility of treating spruce by a sap-displacement process. Sap-displacement treatment takes advantage of the fact that free movement of fluids can occur in undried sapwood; hence if a preservative is forced into the wood tissue it can replace the moisture within provided that the moisture has "a means of escape". A sap-displacement method known as the Boucherie process has been practiced on the Continent for well over a century. In its early form freshly felled poles were placed in rows, usually with their butts raised slightly above their tops, and a water-borne preservative was piped to a sealed cap place over the butt; pressure was obtained by positioning the feed tank about ten metres above the poles (Fig 2.2). In later modifications the process was speeded up by using compressed air from portable cylinders to drive the preservative in. Leachable preservatives such as copper sulphate which were likely to be washed away during the service life of the pole were replaced by more permanent formulations like copper-chrome-arsenate.

A recent improvement developed in Denmark involves the placing of fresh-felled poles in a pressure cylinder. Air-tight caps are placed over the tops of the poles, each cap being connected to a pipe which enables the expelled sap to flow out of the cylinder. Thus when pressure is applied within the cylinder, the preservative is forced into the poles at the butt ends, displacing sap which drains away through the caps and out of the cylinder via the pipes (fig 2.3). This method gives complete penetration of the sapwood and is the most efficient way so far developed for treating spruce poles. It has been found to be effective with either copper-chromate-arsenate or with a water emulsion of creosote.

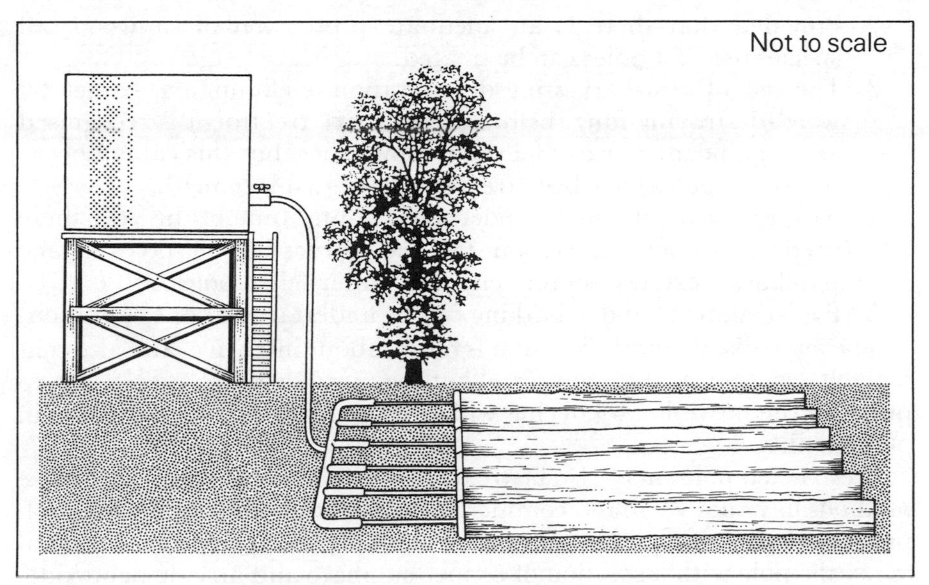

Fig 2.2 Simple sap displacement method for impregnation of fresh-felled roundwood with a preservative.

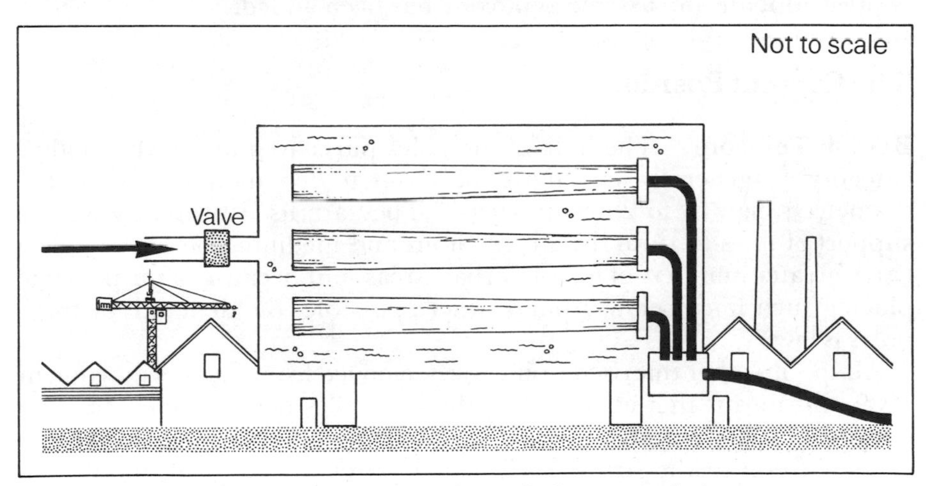

Fig 2.3 Industrial plant for sap displacement method of impregnating fresh-felled roundwood with preservative.

Sap-displacement has a number of advantages over the conventional method of air-drying followed by pressure treatment:

1. The treatment cycle is short because no initial drying is required (drying and fixation of preservative occur simultaneously). Poles have been put into service less than four months after felling improving the cash flow compared with conventional methods.

2. Provided that there is an adequate proportion of sapwood, all species used for poles can be treated.
3. The risk of post-harvesting deterioration is eliminated. Attack by wood-destroying fungi before preservative treatment is recognised as a significant cause of failure in pine poles: but this cannot occur where the poles have been treated by sap-displacement.
4. As the treatment can be undertaken at any time of the year there need be no closed season for the harvesting of poles (some purchasers exercise a preference for winter-felled poles).
5. Pole trimming* and debarking can be undertaken as one operation.

The main disadvantage is that it is more labour intensive, and also that for the less permeable species, "hot-logging" may be necessary to preclude the possibility of drying with a consequent further reduction in permeability.

Fresh felled poles may also be treated by diffusion processes, but these methods have not yet found commercial application in Britain. One such process which has achieved full penetration of the sapwood, has been to spray the pole with a solution of copper sulphate and arsenic pentoxide, followed by a further spraying with ammonia solution to which some copper sulphate and arsenic pentoxide has been added.

The Current Position

British Telecom. The bulk of the poles purchased are in the "light" category** (generally in lengths of seven to ten metres, and in the diameter range 18 to 25 centimetres). They are used primarily for the support of lines to rural dwellings. Stout and medium poles are used to carry greater numbers of lines in urban areas, but because of the policy of placing such lines below ground, where possible, the demand for these poles is low.

All species other than Scots pine were removed from the specification in 1989, this means that almost all of the new poles being erected are now imported from Finland.

In addition to the British Telecom requirements a small number of poles is procured annually by the railways and Hull City Corporation.

Electricity Supply Poles accepted by the Electricity Area Supply Boards are purchased on a "ready for use" basis, which means that they are acquired after drying, trimming, cutting to final shape, boring of bolt

* Pole trimming is an operation in which the poles are rotated against a cutting head to remove any remaining inner bark and snags adjoining the knots.

** BS 1990 makes provision for three size categories light, medium and stout.

holes, and treatment with a preservative. All of these operations are undertaken in the yards of the pole importers. While the decision as to which species to use is one for the individual board, it is largely as a result of this purchasing arrangement that nearly all of the poles in use have been imported from Scandinavia. However, following the joint project between the Forestry Commission and the Electricity Council (referred to above), most home grown conifer species have been acknowledged as suitable for the support of lines of up to 132 kilovolts.

Most of the poles used are of the medium (diameters between 20 and 38 centimetres, at 1.5 metres from the butt end) and the stout categories (diameters between 26.5 and 47 centimetres). These are used mainly for the replacement of poles in rural areas which are at the end of their useful service life. Creosote is the preservative most commonly used, but copper-chrome-arsenate is also acceptable.

Poles for export

A company in the north of Scotland specialises in the preparation of poles for export to developing countries many of which are undertaking extensive rural electrification schemes. The poles have to comply with BS 1990. So far they have been mainly Scots pine treated with creosote.

Other poles

There is a small but locally important demand for round coniferous timber for pole barns, flag poles, rugby goal posts, piling for marine and port construction work and for river and sea defences. The specifications for these varied uses is usually drawn up by the individual architect, engineer or surveyor.

Pole barns are agricultural buildings in which poles are set in the ground to provide the main vertical support. Details of design and construction may be obtained from the Ministry of Agriculture, Fisheries and Food, Block A, Coley Park, Reading, RG1 6DT. Farm buildings generally are covered by British Standard 5502 (relevent part) 1986.

The requirements for strength and service life are similar to those for transmission poles. While straightness is important, many trees in a stand which do not match up to the requirements of, for example, British Telecom, will be acceptable in these other markets.

Pit Props

Some confusion exists in the terms applied to the different types of timber used by the coal mining industry, namely pit props, pitwood, mining timber and sawn mining timber (the latter is dealt with in Chapter 7). Pitwood and mining timber are general terms.

As well as covering square sawn material, the term *sawn mining timber* includes roundwood that has been split by sawing.

Pit props are round (generally coniferous*) timbers, usually peeled and dried (seasoned), used as upright supports in underground workings (Fig. 2.4). At one time they were supplied in a wide range of sizes, from 0.45 m (1.5 feet) × 62.5 mm (2.5 inches) top diameter, to 3.3 m (12 feet) × 250 mm (10 inches); but the bulk of the peeled pit props used outside the South Wales collieries ranged from 62.5 to 200 mm (2.5 to 8 inches) top diameter in lengths of 0.75 to 4.0 m (2.5 to 12 feet). The numbers of sizes of pit props in service today are much more limited, and their use generally has declined in the highly mechanised mines of the 1990s.

The slenderness ratio arrived at by years of practical experience and somewhat limited laboratory tests, which gave good results in the prevention of "sideways buckling", was that the top diameter (under bark) in inches should never be less than the length of the prop in feet, thus establishing a rule of thumb which was easy to remember under the imperial system of measurement.

In South Wales and Monmouthshire much "pitwood", as it was called, was used unseasoned with the bark still on. It was imported from the coastal pine forests of the Landes in south west France and from Portugal, but during and after the Second World War supplies came increasingly from home sources. It was held that pitwood of this type was well suited to the conditions of "squeeze" which were said to be characteristic of the South Wales workings. The imported pitwood was shipped to ports such as Cardiff, Newport, Swansea and Port Talbot, loaded direct from ship to rail, and transported to the colliery yards where it was stock-piled for future cutting into the types and sizes of timber needed underground at any particular time.

Four stock lengths were purchased, each in a range of top diameters:

Length in Feet	Diameter Range in Inches	Proportion by Weight (imported)
4·5	3·5–5·0	15 per cent
6·5	4·0–6·5	45 per cent
9·0	5·0–8·0	30 per cent
13·0	6·0–10·0	10 per cent

Although the use of unpeeled, unseasoned pitwood was widespread in South Wales, up to one third of the pit props used in the late 1930s in this

* A small proportion of the pit props used in the South Wales pits were of unbarked hardwoods. Mines in the Forest of Dean also favoured the use of hardwood props cut locally (which, of course, incurred low transport costs).

region were of the peeled and seasoned kind commonly used in other British coalfields; and the pit prop tests carried out during the 1960s (see below) indicated that there was a case for the greater use of well dried rather than unseasoned pit props. As the final decision as to which materials will be used for underground supports is the prerogative of the individual mine manager, who by law is responsible for the safety of his work force, the change of emphasis from traditional pitwood came but slowly, and some unpeeled pitwood is still in use in South Wales.

The need to have a secure supply of mining timber — especially pit props — on a national basis did much to inspire the formation of the Forestry Commission in 1919 and the initiation of the large planting programmes of conifers after the Second World War.

In the mid 1950s hardwood sawn mining timber became important in roof support techniques (see Chapter 7), greatly boosting the production and use of low-grade hardwood timber, at a time when round softwood

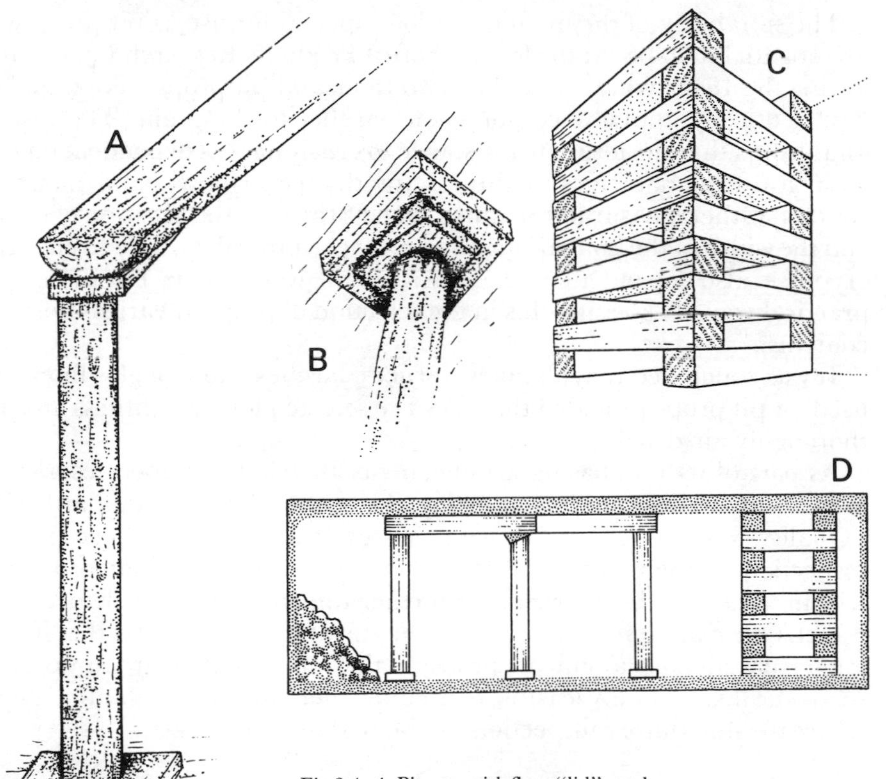

Fig 2.4 *A*, Pitprop with floor "lid", wedge and split; *B*, Pitprop with wedge and lid at roof; *C*, Pillar of chocks; *D*, Pitprops with floor "lids" and splits near working face in pit. Later pillar of chocks replaces pitprops.

timber (mainly pit props) was declining as a result of mechanisation at the coal face (Figs 2.5). By the 1970's the decline in the use of all forms of mining timber had gained momentum, only sawn softwood mining timber remaining relatively stable.

Use of British Mining Timber by the Coal Industry
(000's of cu. metres per year)

	1956–60	1961–65	1966–70	1971–75	1976–80	1981–83	1986–88	1989*
Total	1,058	948	799	541	515	556	371	288*
Round	362	297	208	172	144	137	66	38*
Sawn Softwood	195	160	173	141	131	175	188	250*
Sawn Hardwood	501	491	418	228	240	244	117	

Note 1984–85 figures affected by the miners' strike.
* 1989 British Coal forecast
Sources: 1956–65 G.B. Statements to the 1962 and 1968 Commonwealth Forestry Conferences. Remainder National Coal Board or British Coal.

The suitability of the main coniferous species for use as pit props was investigated in depth at the former Forest Products Research Laboratory during the 1960s when more than two thousand pit props were tested to destruction by applying compression parallel to the grain. The results established that the ultimate resistance to crushing was dependent on two main features, namely straightness (or degree of bow), and moisture content. Other measurable features such as the size and location of knots, and the width of the annual rings had no consistant effect. The strength is, of course, also dependent on the density of the wood tissue, but there is no practical way of assessing this in the selection of props at varied moisture contents.

It was concluded that all pines, spruces, larches and Douglas fir can be used for pit props provided that they are selected for straightness and are thoroughly air-dried.

As part of its purchasing arrangements British Coal operates a strict method of quality control invoking the use of British Standard 5740 "Quality Systems", together with the registration of suppliers and the inspection of their premises. It is therefore imperative that producers wishing to enter this market seek further information from British Coal direct. One consequence of the introduction of this Standard is that it is no longer practicable to cut pit props in the forest, either at stump or at roadside; it is now only feasible to produce them in a well-organised depot where the mandatory inspection can be readily undertaken.

Wood wool

Wood wool is the name given to aggregated strands of wood which are used to protect fragile goods from damage by shock loads. Its other main

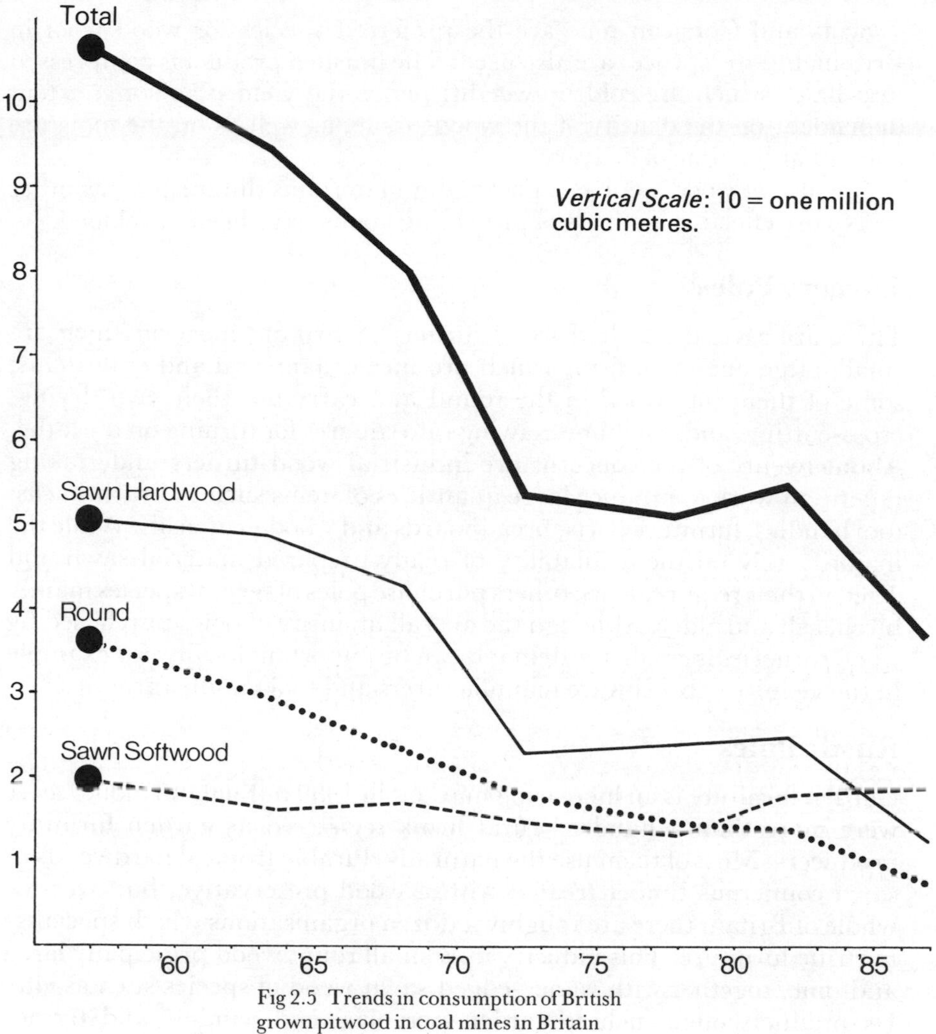

Fig 2.5 Trends in consumption of British grown pitwood in coal mines in Britain 1958 to 1988.

use is in the manufacture of wood wool cement slabs which is described in Chapter 14.

Wood wool is made by incising or scoring a billet of wood to a depth of 0.25 to 0.33 mm at intervals of 3 mm along the entire length parallel to the grain, then planing off the incised wood. Ideally the lengths of the strands should be that of the billet but in practice they are "shortened" by knots and grain deviation.

The "scoring" and "planing" knives are held in a reciprocating mechanism which can process two, three or four billets at a time. The billets must be dried to a moisture content of 20–30 per cent, otherwise there is a risk that the damp fines produced could clog the knives.

Scots and Corsican pine are the preferred species for wood wool in Britain, but the spruces are also used. The finished product is compressed into bales which are sold by weight; hence the yield is to some extent dependent on the density of the wood tissue, as well as on the moisture content at the time of delivery.

The use of wood wool as a packaging material is diminishing as other more cost-effective methods of protecting goods have been developed.

Turnery Poles

There are several hundred wood turners in Britain, most of which are small, often one-man, firms which produce ornamental and craft items; some of them buy wood in the round and carry out their own drying, cross-cutting, and sometimes sawing into squares for turning on the lathe. About twenty of the concerns are industrial wood turners undertaking repetition work to produce large quantities of items such as brush-backs, tool handles, furniture parts, breadboards and wooden utensils. While the majority rely on the availability of ready-prepared material sawn and dried to their requirements, others purchase poles of several species mainly birch, ash and alder. Although the overall quantity of poles purchased by wood turneries is small, the demand can be important locally; for example in the vicinity of brushware manufacturers and tool handle turneries.

Rustic Poles

Garden furniture is an increasing market; in 1989 in England alone, there were more than a hundred rural firms registered as garden furniture producers. Most of them use the naturally durable tropical hardwoods or sawn coniferous timber treated with a wood preservative, but over the whole of Britain there are roughly a dozen organisations which specialise in rustic furniture. This industry uses small roundwood principally larch and pine, together with waney-edged sawn wood of species such as elm. Its products often include garden pergolas, and window and terrace flower boxes. Garden centres which commonly retail these commodities may also sell the poles on the Do-it-Yourself principle.

Walking Sticks

Walking sticks, particularly for use by hospitals, are made by a few specialist firms from young, usually three year old, chestnut coppice. Additionally, there are several small organisations making "leisure" walking sticks, often with carved handles, from ash, chestnut and hazel, as well as from imported materials such as cane.

BS 5181: 1975 (1988) covers the specification for wooden walking sticks, light, medium and heavy, in lengths of 915, 965 and 1015 mm.

3

Sawlogs and their Conversion to Sawn Timber

Log Grading

Sawlogs are marketed in a variety of qualities and dimensions ranging from those capable of yielding only lower value products such as sawn mining timber and dunnage, to stout high grade logs likely to produce a good out-turn of premium commodities such as furniture components, coffin boards and joinery timber.

Although for some purposes the *minimum* size of hardwood sawlog that is acceptable is around 30–40 cm diameter in lengths of 2.5 metres, the *average* size which it is economic for a hardwood sawmill to handle will be nearer five metres in length, with a minimum top diameter of 50 cm underbark.

Softwood sawlogs are commonly sold in lengths of at least three metres with a minimum top diameter of 14 cm underbark. But a sawmill supplying a general range of softwood sawn lumber would require an average log size of five metres in length by 30 cm top diameter underbark.

Sawmills engaged in the continuous production of softwood fencing panels use specialised machinery. The log sizes which they purchase are given in Chapter 6; the lengths are determined by the preponderence of sawn pieces needed to make standard six foot (1.8 metres) fencing panels.

There are a few sawmills that undertake the supply of longer lengths of softwoods or hardwoods for such purposes as dock piling, the restoration of historic buildings, jetties, river and sea defences, boat-building, and the construction of large farm buildings and warehouses. For these purposes logs up to 20 metres in length with a minimum top diameter of 30 cm are required.

Formal log grading is undertaken on a relatively small scale in Britain. This is partly because many sawmillers purchase their timber standing, and partly because of the role played by round timber merchants who, through an intimate knowledge of the needs of individual sawmillers and other consumers, are able to channel roundwood of the required size and quality to the most remunerative market, without reference to published grading rules.

Log grading should be regarded solely as an aid to marketing, especially where the lots advertised are for sale by auction or tender. A grading system has, therefore, two fundamental requirements. It must incorporate sufficient precision to enable a potential purchaser to make a bid without the necessity of an inspection, and it must be cheap to operate. If a grading system involves the use of measuring equipment and requires more than a few seconds per log, it will only be suitable for the more valuable species. For the general run of softwood sawlogs, any log grading system that is to find wide acceptance should be designed so that a trained worker can make an instant assessment of the grade. Attempts to devise grading rules in the 1960s were unsuccessful because they involved time-consuming measurement of the log's features. One such system was compiled by a committee drawn from merchants and growers, and was serviced by the former Forest Products Research Laboratory, which published the rules; while another method, created by the late Dr S. E. Wilson, although easy to memorise,* gave a disappointingly low correlation with the yield of sawn wood.

Concurrently the East (England) Conservancy of the Forestry Commission was operating a system of softwood log classification by size of log and incidence of defects; this incorporated an "elite" category for logs which had received early pruning. The innovator Mr R. Chard subsequently introduced the system into the South (Scotland) Conservancy.

More than a decade later the Forestry Commission, after consultation with the British Timber Merchants' Association and the Home Timber Merchants' Association of Scotland, published a system of softwood log classification which was revised in 1989. Details are given in the Appendix.

Straightness is of paramount importance in the selection of sawlogs, consequently buyers employed by modern sawmills with automated equipment have devised a method of selection which gives consistantly high yields. This involves inspecting the top end to determine the largest square which could be cut from the log, then "projecting" this square along the entire length. If the projected square is deemed to have "broken" the surface at any point, the log is rejected.

---

* The Wilson system graded softwood logs by straightness and the diameter of the largest visible knot; viz A = arrow straight, B = bent in one place only, and C = crooked — namely bent in two planes. Thus a "B3K1½" log would have a deviation in one plane only of three inches (75 mm), and a maximum knot size of one and a half inches (38 mm).

Apart from straightness, taper, freedom from fluting, and absence of decay and shakes* are also important. Where decay, which is mainly butt rot caused by the fungus *Heterobasidion annosus* (formerly *Fomes annosus*), fluting or butt sweep (ie a localised distortion at the base of the stem) occur, it is common practice to remove a short length from the bottom of the log for conversion into pallet boards or fencing material, leaving a second length of higher quality.

In coniferous timber knot size has a considerable influence on log grade. In well-managed crops under British conditions, knot size is mainly dependent on species; hence an experienced purchaser of softwood logs will know that the incidence of large knots is much lower in the larches and spruces than in the pines and Douglas fir. Knot type is, for some uses, as important as knot size; thus small loose knots encased in bark can make timber unsuitable for joinery since they fall out in drying. This defect spoils much larch that would otherwise make high quality joinery wood.

There are also international systems of log grading. In 1966 the Timber Division of the Economic Commission for Europe (ECE), and the European Office of the United Nation's Food and Agriculture Organisation (FAO) (now the ECE/FAO Agriculture and Timber Division) drafted softwood grading rules to facilitate trade between member nations. The task proved to be more complicated than initially envisaged despite the existence of national rules. Eventually in 1968 the whole subject of log grading was referred to the Timber Committee of the International Organisation for Standardisation (ISO), but little progress has been made due mainly to lack of interest by the trade in member countries. However, by 1988 agreement was reached on the classification of visible defects, ISO 4473: 1988 (E). See pages 46/47.

Within the Common Market a Directive, number 68/69 "Measurement of Wood in the Rough" dated 1st July 1973 was issued with the objective of standardising measurement practices and size categories within the European Economic Community, but it does not affect British practice significantly because only veneer logs are exported from Britain to the EEC in appreciable quantities.

Pruning for the improvement of sawlogs

The high pruning of selected coniferous trees has been a controversial issue in plantation management. To be effective the operation must be undertaken early in the life of the crop, preferably at least three decades before harvesting.

Because of the long interval before the original investment plus the

* Shakes are splits or fissures within the growing tree seen on the butt end of the felled log. They are prevalent in some hardwoods eg sweet chestnut and oak.

Table 3·1 Coniferous and broadleaved sawlogs – visible defects – classification ISO 4473

Classification of Defects*

Group	Subgroup	Variety	
1 Knot	1.1 Flush knot	1.1.1 Sound knot: decay knot	
		1.1.2 Unsound knot	
		1.1.3 Rotten knot	
	1.2 Overgrown protruding knot: burl		
2 Shake	2.1 End shake	2.1.1 Heart shake	2.1.1.1 Simple heart shake
		2.1.2 Ring shake	2.1.1.2 Compound (Star) heart shake
	2.2 Side shake	2.2.1 Frost crack and shake caused by lightning	
		2.2.2 Drying shake according to depth	
		2.2.3 Shallow shake	
		2.2.4 Deep shake	
		2.2.5 Through shake	
3 Defects of trunk shape	3.1 Curvature	3.1.1 Simple curvature	
		3.1.2 Compound curvature	
	3.2 Knob		
	3.3 Root swelling: buttress	3.3.1 Round root swelling	
		3.3.2 Veined root swelling	
	3.4 Ovality		
	3.5 Tapering		
4 Defects of structure	4.1 Slope of grain		
	4.2 Reaction wood		
	4.3 Double or multiple pith		
	4.4 Removed pith		
	4.5 Scar		

	4.6 Inbark	4.6.1 Opened inbark 4.6.2 Closed inbark	
	4.7 Canker		
	4.8 False heartwood**		
	4.9 Heart sapwood		
5 Defects caused by fungi	5.1 Fungal heartwood stains and streaks		
	5.2 Fungal sap coloration	5.2.1 Blue stain 5.2.2 Coloured sap stain	
	5.3 Suffocated wood**		
	5.4 Rot	5.4.1 Sap rot	
	5.5 Hollow	5.4.2 Heartwood rot	
6 Damage	6.1 Damage caused by insects (insect holes)	According to depth 6.1.1 Surface insect hole 6.1.2 Shallow insect hole 6.1.3 Deep insect hole	According to diameter 6.1.3.1 Small insect hole 6.1.3.2 Large insect hole
	6.2 Damage caused by parasitic plants		
	6.3 Bird holes		
	6.4 Alien inclusion		
	6.5 Char		
	6.6 Mechanical damage	6.6.1 Bark shelling 6.6.2 Blaze 6.6.3 Incision 6.6.4 Saw-cut 6.6.5 Off-chip 6.6.6 Shear 6.6.7 Extraction	

* For definitions of the defects, see ISO4474: 1988, Coniferous and broadleaved tree sawlogs – Visible defects – Terms and definitions

** The defect is typical only of broadleaved sawlogs

accrued interest can be recouped it is an expensive operation. Moreover it carries an element of risk, partly because the premium markets envisaged for the final crop might not exist thirty years hence (for example it might be unwise to prune larch in anticipation of a future market for boatskin timber), and partly because the stand could be damaged or destroyed by windthrow before much clear timber had been added to the stems.

Pruning should normally be restricted to those species such as Scots pine, Corsican pine and Douglas fir where a relatively high proportion of the branches exceed 25 mm in diameter; in hardwoods there is usually a good case for pruning poplar, particularly if the production of peeler logs for the vegetable crate industry is envisaged. However, the removal of live branches should not be so intensive as to result in an appreciable reduction in the stem increment. It should also be realised at the out-set that it is only the outer layers of boards which can be expected to attract an enhanced price and that a proportion of clear wood resulting from the pruning is "wasted" as slabwood and sawdust.

On the other hand pruning brings tangible benefits other than the up-grading of the sawn out-turn. Harvesting costs can be reduced because there is less "hanging" of trees during felling, and there are fewer branches to remove in the delimbing operation. Additionally there is the unquantifiable benefit of the improvement of the aesthetic appeal of the crop; this is particularly true for species such as Scots pine which display much more attractive stems when no dead branches are present.

Conversion to sawn timber

Prior to the nineteenth century the cutting of round timber to square sections was done manually, either by the use of pit saws (Fig 3.1A) — a method still widely used in Third World countries — or, when only one piece was needed, it could be trimmed down to the required dimensions by adze.

An alternative method to sawing sometimes undertaken in those species such as oak, ash and sweet chestnut, which cleave readily, was to split the wood by driving wedges along the grain. Afterwards the rough sections thus produced could be adzed to the final shape or further split with a knife known as a froe. Smaller sections could be turned on a pole lathe. Today in the developed world most logs are broken down in sawmills although occasionally a chain-saw attachment may be used for small logs (Fig 3.1B).

In modern sawmilling practice it is usual to remove the bark before the first cut is made. This reduces the wear of the cutting edges caused by grit lodged in the bark; it also ensures that the slabs produced as a by-product are free from bark and thus suitable as a raw material for the higher grades

Fig 3.1 *A*, Pit sawing.
B, Chain saw attachment for through and through sawing of small logs.

of pulp, and as the preferred material for some of the panel product industries. The bark itself can be further processed and sold for a wide range of applications (see Chapter 12). Another and more recent operation before the log actually enters the sawmilling process, is for any butt swelling to be removed by a machine commonly called a butt-reducer.

Recent decades have seen the fully mechanised conversion to

square-edged timber by circular saws, reciprocating frame saws, band saws and profile chippers. In sawmilling, the proper initial presentation of the log to the saw is essential for maximum profitability. To achieve this automatic electronic scanners or lasers, coupled to computers which have been programmed with information on current markets, have been installed at some sawmills; others continue to rely on the experience of the operator of the first saw.

Portable sawmills have been in use for many years. A traditional Scottish sawmill of the type erected in the woods for use over periods of up to two or three years and truly portable modern sawmills are shown in Figs 3.2 and 3.3

Fig 3.2 Semi-portable circular sawmill ("Scotch bench") in common use, especially in Scotland, up to the 1950s. Occasionally found up to the 1970s. Courtesy Ronald D. Gordan.

All types of sawmilling other than profile chipping have been in continuous use throughout this century. Although during this period each type has undergone substantial technical development, the broad advantages and disadvantages of each remain. The pros and cons of today's systems are considered below.

Circular saws

Circular saws (Fig 3.2) have the advantages of low capital outlay, relative ease of maintenance, and flexibility in operation in that they can be used for cant,* through-and-through or quarter sawing (see page 55 and Figs 3.6 and 3.7). The maximum depth of cut is, of course dependent on the diameter of the blade, and where large diameter blades are fitted there can be some loss of accuracy, unless thicker blades are used with a consequent increase in wastage. Some manufacturers incorporate a system of water sprays to cool fast rotating blades.

Inserted tooth saws, in which the individual teeth can be removed for sharpening or replacement, have the advantage of ease of maintenance, but produce kerfs of the order of 6 mm (¼ inch) and over, giving a high wastage of timber.

Frame saws

Frame saws are of two different types. The first comprises a single blade mounted horizontally in a frame, and in Britain was formerly used to break down hardwood logs. The second type, consisting of several blades vertically mounted in a reciprocating frame (Fig 3.4), is more common in countries with a larger supply of uniform small diameter logs than Britain has. At one pass they can produce an unedged cant and a number of side boards, which can be further sawn by twin or single circular saws to produce square-edged lumber. They are somewhat inflexible because they cannot produce some of the cutting patterns described below, which usually give a more valuable sawn out-turn than can be obtained by through-and-through sawing. The blades have to withstand an abrupt change in direction inherent in a reciprocating mechanism, and thus, to withstand buckling, have to be thicker than the blades of band or circular saws, resulting in more waste in the form of sawdust. On the other hand they have a well-earned reputation for accuracy.

The distance at which the blades are set depends on the market's requirements for different thicknesses of timber, and on the diameter and taper of the logs to be sawn. It is obviously uneconomic to have to stop the saws to reset the blades during a production run. As freshly sharpened

* A cant is defined as a sawlog with one or more slabs removed.

Fig 3.3a A modern portable sawmill. Courtesy, Forestor

Fig 3.3b Tractor driven modern portable vertical band saw. Courtesy, Forestor

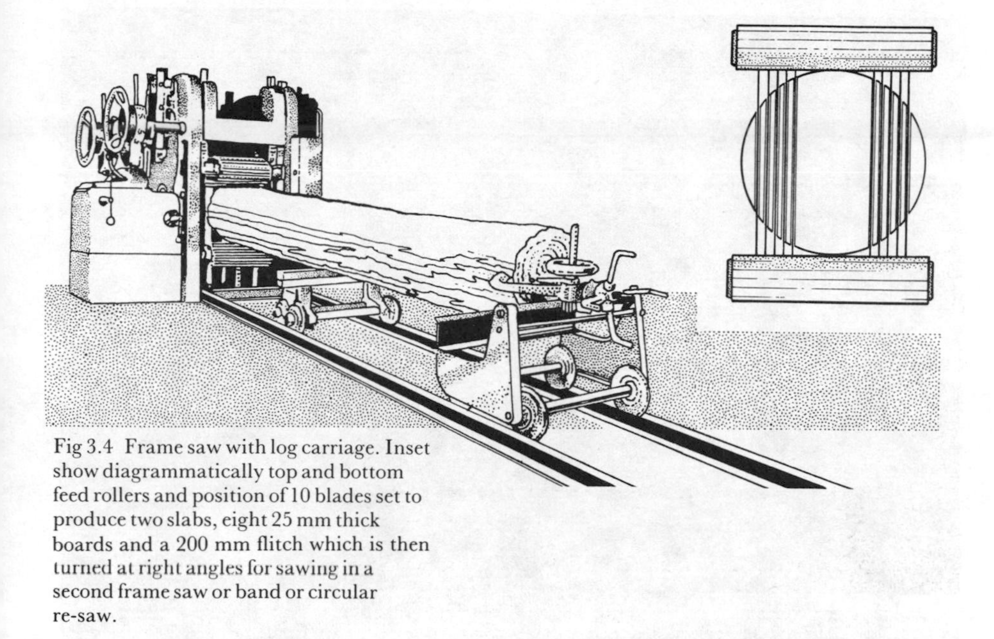

Fig 3.4 Frame saw with log carriage. Inset show diagrammatically top and bottom feed rollers and position of 10 blades set to produce two slabs, eight 25 mm thick boards and a 200 mm flitch which is then turned at right angles for sawing in a second frame saw or band or circular re-saw.

blades are needed roughly every four hours, re-setting is best undertaken during lunch breaks and between shifts. This necessitates the pre-sorting of logs to ensure that there will be enough of one size category for four hours sawing.

Band saws

Horizontal or vertical band saws (Fig 3.5) have been used in Britain for the cutting of hardwoods for most of this century. Since they can convert large diameter logs more efficiently than circular saws, they are invariably used for the initial sawing of stouter boles. Their use in both hardwood and softwood mills has increased substantially since they produce not only a narrow kerf (frequently less than 2 mm), but are capable of cutting at high speeds. There is a risk that cutting accuracy may be sacrificed for speed.

Band saws have always been more difficult to maintain than circular saws, but modern automatic sharpening and setting devices have replaced much tedious hand work.

Profile chippers

Profile chippers are in essence rotary planing machines which reduce the rounded edge of the log to a flat surface. They are usually arranged in pairs so that when a log is passed between them a two-sided cant is

Fig 3.5 Vertical bandsaw.
Courtesy Stenner of Tiverton Ltd.

produced. In some systems this cant is conveyed to conventional saws for further conversion and in others it is turned through a right angle and recirculated between the profile chippers to produce a square cant, which is then converted further by sawing.

Profile chippers can be effectively integrated with pulp mills or wood chipboard mills which utilise the chippings or flakes produced. The advantage of this system is that little waste is produced in the form of sawdust; a disadvantage is that potentially high grade timber on the outer part of the log (i.e. free from juvenile wood, knots and exposed pith) is reduced to chips — which will probably attract a lower return.

The advantages and disadvantages of the various types of saw are listed in the table on page 57.

Cutting patterns

Some typical cutting patterns are shown in Figs 3.6 and 3.7. To appreciate the reason for converting sawlogs in a particular way, it is worth considering the cross-section of typical plantation-grown coniferous sawlog, and to examine the effect of its gross features on the grade of wood produced.

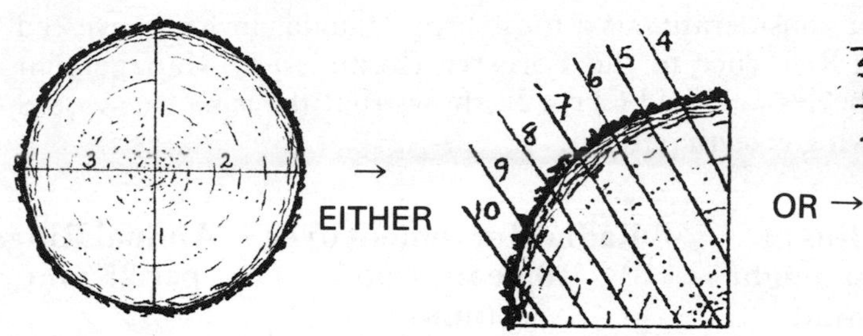

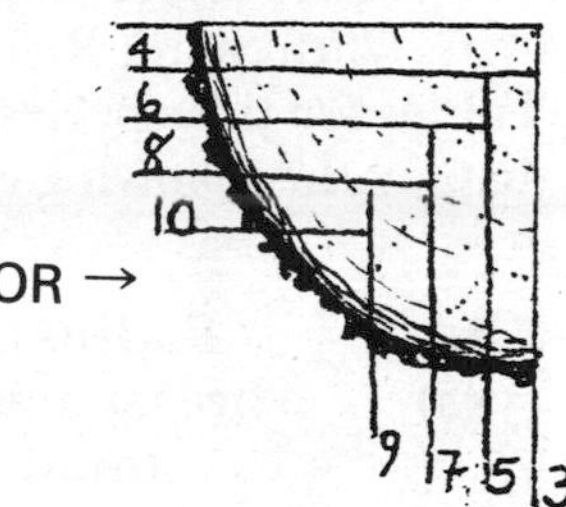

Quarter sawing seqence of cuts

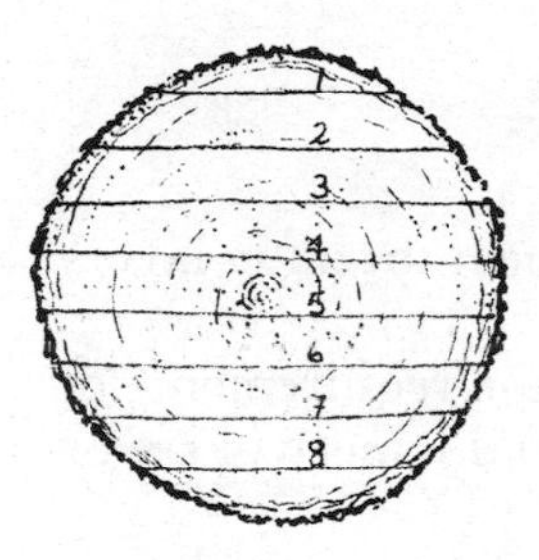

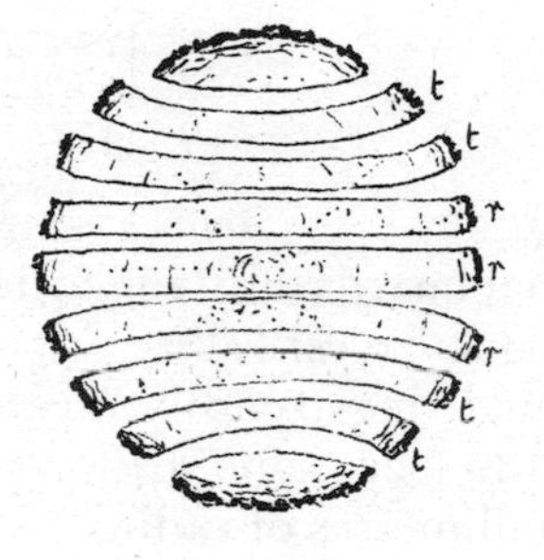

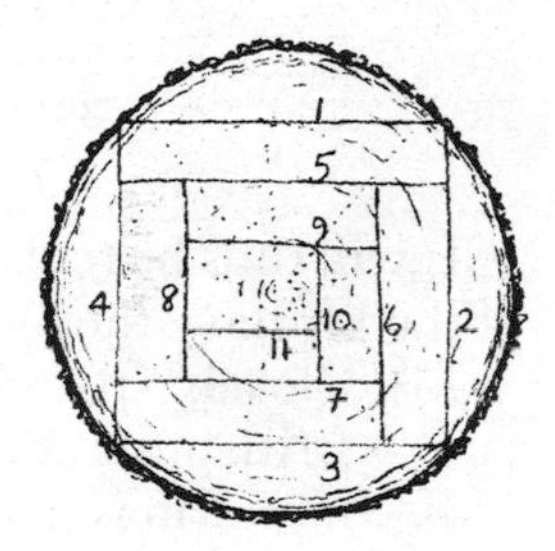

Through and through sawing sequence of cuts

t = tangential piece liable to cupping
r = radial (quarter sawn) little distortion

cant sawing sequence of cuts

Fig 3.6 Typical sawing patterns — Circular or Band sawmill.

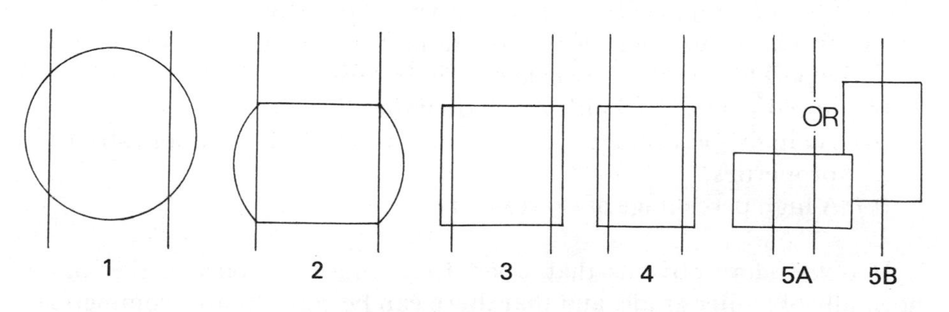

Fig 3.7 Profile chipper cuts 1 and 2. Twin band saws cuts 3 and 4. Cuts 5A twin band saw; or re-saw for cut 5B.

The case under consideration is a main crop Douglas fir log harvested in its 70th year. Reference to the Forestry Commission Management Tables for this species for Yield Class 22 shows that the *mean* tree would display the following growth pattern:

Age (yrs)	Radius at breast height (mm)	Radial Increment over 10 year period (mm)	Annual Rings per 25 mm
10	46	46	5
20	85	39	6
30	152	67	3–4
40	211	59	4
50	258	47	5
60	298	40	6
70	331	33	8

It will thus be realised that boards cut tangentially near the outer layers of the log have the following features:

(a) Closely spaced annual rings readily meeting the requirements of British Standard 1186 (see page 96) for external joinery of six or more rings per 25 millimetres of radius.
(b) Knots, where present, passing through the thickness of the board giving a complete absence of splay knots.
(c) Absence of unsightly pith.
(d) Absence of juvenile wood with inferior strength properties.
(e) A high proportion of sapwood (in many instances 100 per cent), which may be undesirable for some purposes, but is more readily treated with preservatives.

Wood cut from the inner layers of the log, on the other hand, has these features:

(a) Widely spaced annual rings making it unacceptable for external joinery, and in extreme cases for structural work.
(b) A relatively high incidence of knots many of which are of the splay type which seriously weakens thin boards.
(c) Streaks of exposed pith which disfigure the surface.
(d) A high percentage of juvenile wood with inferior strength properties.
(e) A high percentage of heartwood.

It is therefore obvious that wood from the outer part of the log is normally of higher grade, and that there can be considerable commercial benefit in optimising the yield of such wood. This is achieved by cant sawing in which three or four slabs are first removed for resawing into high grade boards which depending on the log taper may be somewhat short in length. After this further longer high grade boards can be cut from the first cut of all four sides of the cant. If the log is known to have come from a tree which has received early pruning, cant sawing is essential to optimise the yield of clear timber.

The cutting of narrow boards from the vicinity of the pith is to be avoided, because knots positioned along the rays appear as splay knots and are a serious cause of weakness.

There are other methods (Fig. 3.6) of sawmilling seldom practiced for British softwoods, but should be mentioned for completeness. Rift or quarter sawing is a cutting pattern which maximises the quantity of wood cut in a radial direction. It finds particular application for those ornamental hardwoods such as oak and London plane where the appearance can be enhanced by exposing the figure given by the rays. It can also be employed to increase the wearing properties of softwoods. For example if Douglas fir or pine used for an industrial floor is cut tangetially there could be a tendency for the growth rings to "shell-out" and the floor to become splintery, but if rift sawing is used so that the radial surfaces appear on the surface of the boards the wearing properties would be improved considerably.

Another method of specialised sawing is to makc thc cuts parallel to the bark rather than parallel to the pith. This results in longer lengths of straight-grained higher grade higher value wood than obtained by cant sawing. However, the sawn out-turn is lower because the centre of the log becomes wedge shaped and has limited uses. Wood cut parallel to the bark can be used for such exacting purposes as boat-skins, ladder poles and banister rails where straightness of grain is important.

	Circular	**Frame**	**Band**	**Profile**
Capital Cost	Low	Moderately High	Moderately High	High
Maintenance requirements	Low	Moderate	Moderately High	High
Flexibility	Good	Inflexible	Good	Inflexible
Wastage	Moderate	High	Low	Low*
Accuracy	Moderately Good	Good	Moderate	Moderate

* Much of the high grade timber from the outer part of the log is reduced to chips.

4

The Drying of Wood

The reasons for drying

Timber felled at any time of the year has a high moisture content. In low density species which have a high proportion of sapwood, such as European spruce, the initial moisture content can be as high as 200 per cent of the oven dry weight, but in medium density hardwoods like beech it is usually below 100 per cent. There is, of course, substantial variability within the species, but as the proportion of sapwood (expressed as mean radial depth) is closely related to the biomass of the crown; trees exhibiting large green crowns tend to have timber with higher moisutre contents than those with short narrow crowns. It is sometimes held that wood in the standing tree is appreciably drier during the winter months because "the sap is down"* or because "the sap has not yet started to rise", and for this reason winter felling should be preferred, implying that winter-felled wood is less troublesome in drying than summer-felled wood. While it is accepted that there is some seasonal variation in the moisture content, and especially in the tension to which the water in the sapwood is subjected, timber felled in the winter months is wet and heavy, and it should be dried in the same way as that felled at other times of the year. Differences in the performance of winter and summer felled wood are attributable to external factors such as the air temperature, the relative humidity and the activity of pests.

The moisture content of wood is one of the most important factors in its

* Oak bark for the manufacture of high quality leather is harvested during "the sap peeling season" or "when the sap is rising" namely during the late spring and early summer weeks. This is unconnected with the movement of moisture in the stem. The relative ease with which the bark can be stripped at this time of the year results from the condition of the cambium underlying the bark which is then dividing most actively.

performance in use, and for most purposes some drying needs to be undertaken to ensure satisfactory service.

The advantages of dried wood are:

1. Most strength properties are increased significantly. For example, the compressive strength of European spruce (parallel to the grain) is more than doubled on drying.
2. Dried wood is less likely to shrink and distort (this depends upon the moisture content to which the wood is dried).
3. Dried wood is easier to work and finish with machine and hand tools.
4. It takes paint and varnish better.
5. Attack by pinhole borers (Ambrosia beetles) does not occur.
6. It is less prone to attack by surface moulds, wood destroying or staining fungi, and some insect pests.
7. It usually cleaves more readily.
8. It is much less resistant to penetration by wood preservatives and fire retardants (except where a diffusion or sap-displacement process is used).
9. Corrosion of metals is less likely.
10. It is much lighter in weight.
11. It is more pleasant to handle.
12. Dampness does not spread to materials in contact with the wood.

On the other hand undried wood has a few advantages:

1. It is less prone to splitting when nailed, although some splits may develop when further drying occurs.
2. In some species like elm the resistance to impact loading can be more than ten per cent greater.
3. Nails driven into green wood often "rust in" giving them a much greater resistance to withdrawal. Pallet manufacturers may prefer partially dried wood for this reason. (Where nails shaped to resist withdrawal are used "rusting in" may not be so important a factor).
4. The power consumption of cutting machinery is lower.

How Timber Dries

In the fresh-felled state, where up to two thirds of the mass can be moisture, the sapwood is fully saturated with its cell cavities completely occupied with water, and heartwood nearly so. As drying proceeds to around 28 per cent of the oven dry weight, the water in the cell cavities is evaporated away, but a substantial amount of moisture is retained within the cell walls. This level, above which no appreciable shrinkage occurs, is known as the fibre saturation point (it is a hypothetical concept because moisture is unevenly distributed within the tissue).

Subsequent drying therefore involves the removal of the water retained in the cell walls which shrink significantly as a result. Such shrinkage becomes apparent in round timber as surface checks and splits accompanied by a small loss in diameter. In sawn wood distortion may result from the shrinkage in addition to possible surface checks (Fig 4.1).

While the colloquial term for drying, namely seasoning, probably had its origin in the exposure of wood to all four seasons, very little if any drying occurs during the late autumn and winter months (some species can actually pick up moisture during this period), and the most rapid drying takes place in the late spring and early summer. Drying tends to take longer during the late summer months when atmospheric humidities are higher. In the British climate natural or air-drying usually brings the moisture content down to about twenty per cent and artificial methods should be employed if lower levels are required. Nevertheless, moisture contents as low as twelve per cent have been recorded in sawn softwood in outdoor situations in mid-summer. Artificial drying down to a moisture content of eight per cent is sometimes necessary, but it is not usually taken below this level because, for example, it would result in a considerable increase in the power requirements for cutting and machining.

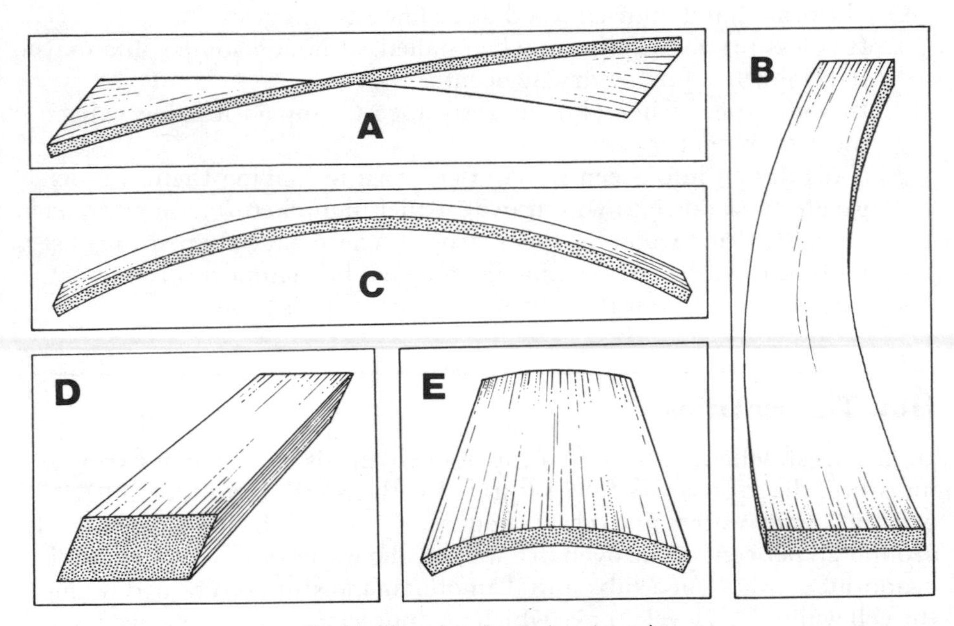

Fig 4.1 Distortion after drying, *A* = twist, *B* = spring, *C* = bow, *D* = diamonding, *E* = cup

The drying of round timber

While it is unwise to allow sawlogs to dry before conversion because the development of surface checks can reduce the value of the sawn out-turn*, drying is required for most applications when timber is used in the round; these include pit-props, fencing (other than temporary fencing), and telegraph and power transmission poles. There have been a number of studies on the drying of roundwood at stump and in the forest depot which have yielded much useful information on the effect of bark removal, log diameter, and season of felling on the rate of moisture loss, as well as on the incidence of fungal attack and the apparent deterioration of the wood. Removal of bark roughly doubles the rate of drying, but as one might expect the drying of unbarked roundwood is more rapid in species such as spruce which have a thin bark than in thick barked species. Thorough de-barking is, therefore, a prime requirement in the efficient drying of roundwood. After this has been done, the poles or stakes should be cross-piled well clear of the ground to facilitate good air circulation. Half-round timber should be cross-piled in the same way, but with the curved surface uppermost, to enable rain water to run off. Old tram rails or second hand rolled steel joists make a good foundation for the stack (Fig 4.2). Roofing the stack is usually unnecessary in the British climate.

Artificial methods, namely the use of energy to expedite drying, are seldom applied to roundwood in the UK, because moisture contents below 22 per cent are not normally required. A moisture content at this level is quite adequate for preservative treatment by conventional methods. The most notable exception has been the application of dehumidifiers to dry fence posts in the high rainfall areas of Ireland.

The air-drying of sawn wood

The stack for air-drying of sawn wood should be erected clear of the ground, preferably on an open weed-free site (Fig 4.2). Here too, old rails or rolled steel joists may be used for the foundation.

Cross-piling is not suitable because for best results contact between the individual pieces should be avoided; this is achieved by the use of alternate rows of spacing slats known as "stickers". Stickers should be of a size to give a space between the rows of boards of between 15 mm and 25 mm. The horizontal gaps between the boards should be a minimum of 20 mm. If the timber being dried is of high value it can be beneficial to place a roof over the stack in areas of high rainfall.

If stickers are not available, a method known as pigeon-hole stacking can be used. For this the first layer of boards is arranged at wide spacing,

* Some merchants undertake the initial drying of oak before the logs are converted, accepting deterioration of the less valuable sapwood.

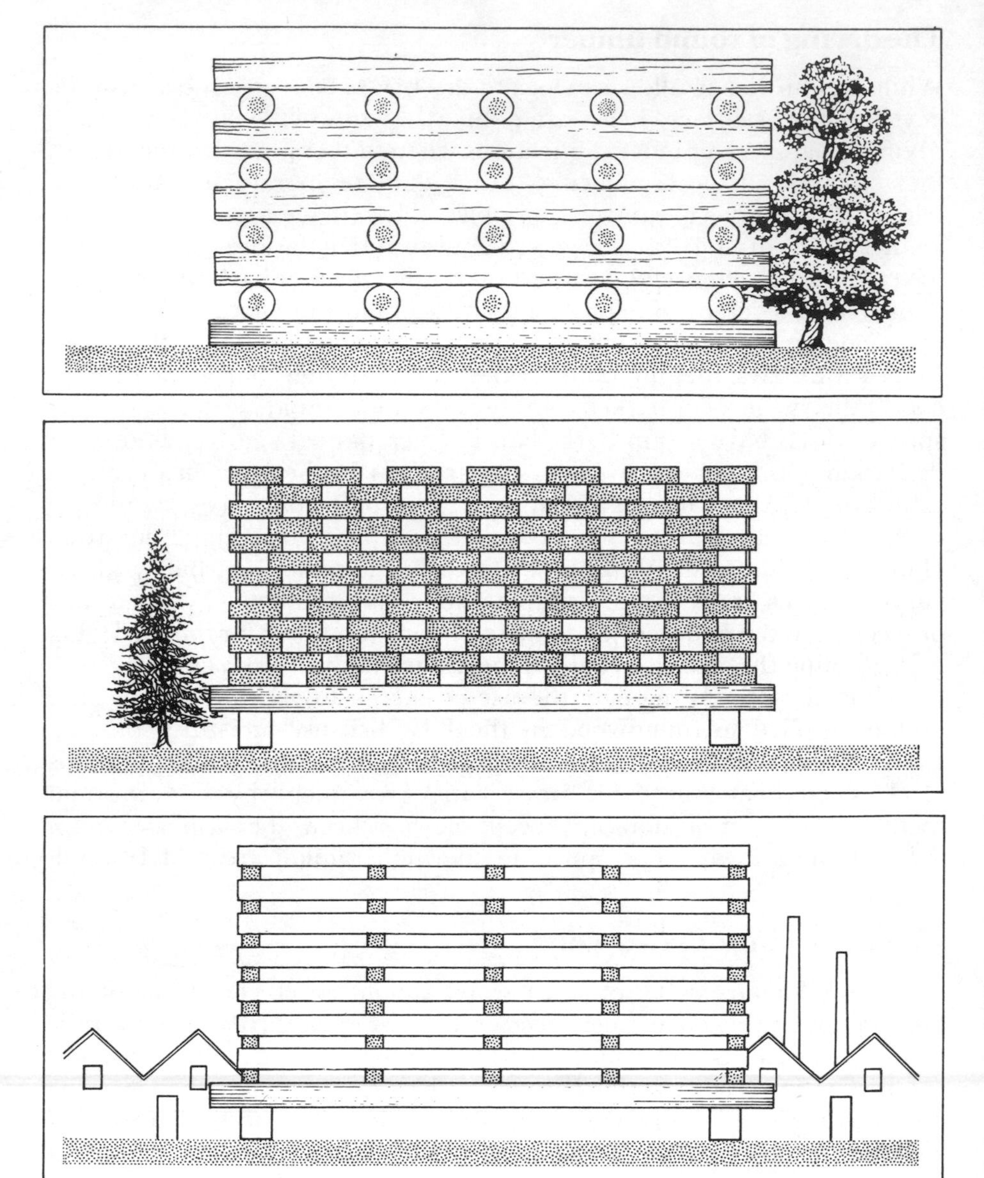

Fig 4.2 Stacking of round and sawn timber to dry. Pigeon-hole stacking in central drawing.

preferably along the direction of the prevailing wind, and the second layer is placed so as to cover the gaps between the boards of the first layer and so on.

While this system is better than close-piling it is less efficient than piling with stickers because the wettest part of the wood (the sapwood) is the least well ventilated. (Fig 4.2).

Some species such as sycamore can suffer a serious loss in value because wood staining fungi may develop in the vicinity of the stickers. This risk can be reduced substantially if the boards are stacked vertically for initial drying.

Although economic pressures may require stacking to be undertaken continuously at all times of the year, the best results are obtained in late winter so that full advantage can be taken of the rapid drying which occurs during the spring months (Fig 4.3).

Air-drying has the obvious disadvantage that very little can be done to control the rate at which it occurs, and losses can be caused by unpredictable spells of weather. For example, wet, warm, humid spells can encourage and will favour the growth of mould, stain or even wood destroying fungi (rotting fungi present in an incipient form in the growing tree can continue to develop); while periods of hot dry weather can cause surface checks and end splits to appear. Even so, it remains the most economical method of drying timber for use out-of-doors and for many internal structural purposes; but it is not adequate for furniture, joinery and flooring where subsequent drying in service can lead to excessive shrinkage and distortion.

Kiln drying

The application of energy as heat and forced air circulation within an enclosed kiln is essential to achieve uniform drying of batches of wood to specified moisture levels. As already noted, air-drying normally attains moisture contents around twenty per cent, but where wood is to be used in modern housing with central heating, double glazing and roof and wall insulation, much lower levels are required. If kiln drying is used, the temperature, humidity and velocity of the circulating air can be controlled, and the uniformity of the dried timber at a specified moisture content virtually guaranteed.

The procedure is to open pile the wood using stickers in a well insulated kiln, using initially circulating air at a high relative humidity and an elevated temperature, and to subsequently adjust these to control the rate of drying. If the temperature is too high, collapse of wood tissue can occur in low density species; and if the drying is allowed to proceed too quickly, not only are surface checks and end splits likely to develop, but also another defect, called "case hardening", may occur, in which inadequately dried wood becomes enveloped in dry wood resulting in troublesome internal stresses. Yet a high initial temperature is useful to ensure the rapid demise of any harmful fungi or insects.

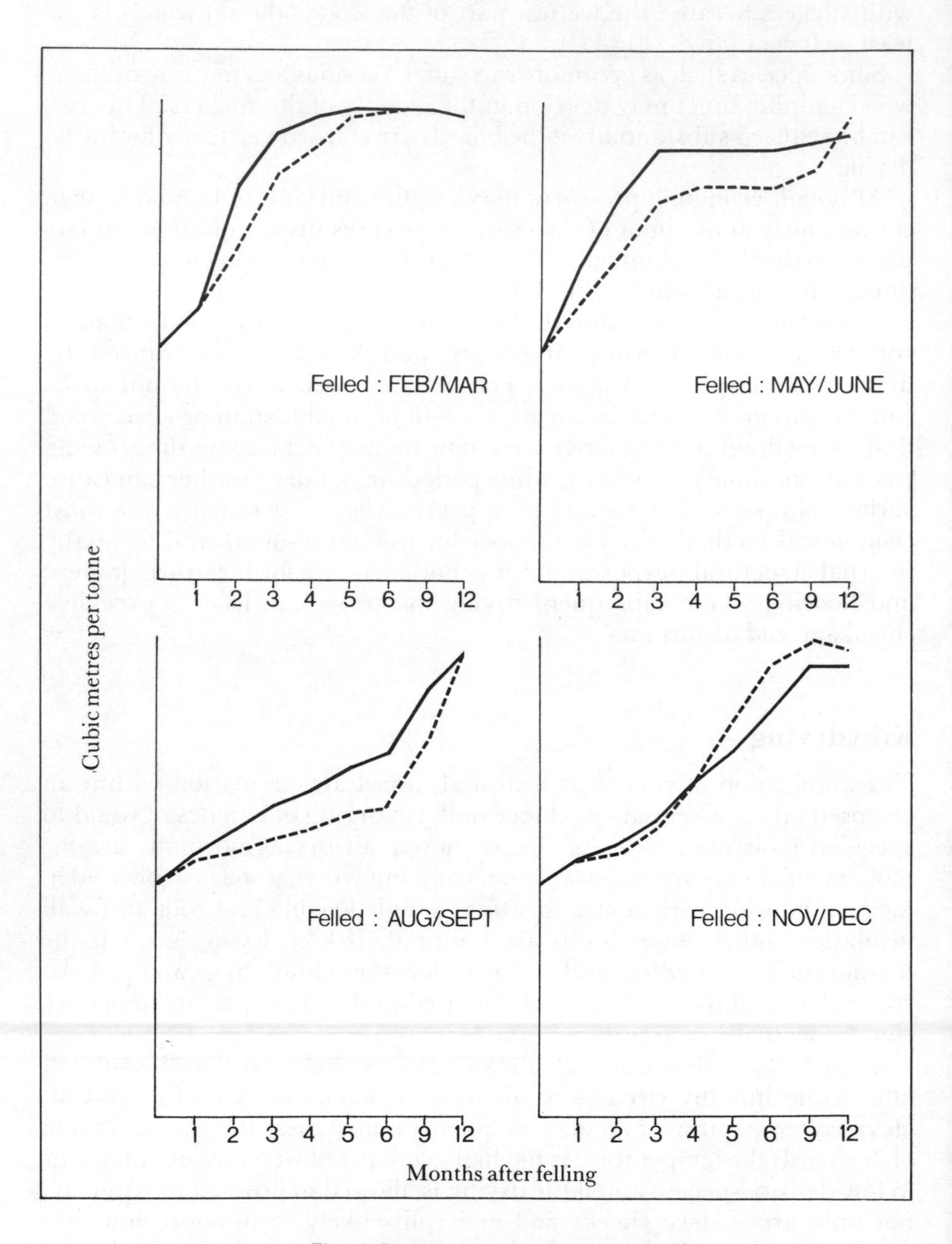

Fig 4.3 Rate of drying of peeled (debarked) roundwood felled at different times of the year. Poles of 10 cm DBH (——). Poles of 16 cm DBH (------)

Techniques have been evolved recently to expedite the drying of softwoods by using higher temperatures, but these might cause problems in machine stress grading, because there are some indications that the relationship between stiffness and ultimate bending strength (on which this method of grading depends) could be confounded. Furthermore, the darkening effect produced by high temperature drying, especially in the paler species, may be disliked for some uses.

Experience has shown that some of the more difficult hardwoods such as oak intended for furniture manufacture are best dried by a combination of kilning and air-drying. The wood is first kiln dried down to fibre saturation point, after which it is placed in the open for air-drying, and then finally kiln dried to the specified moisture content.

Other methods of drying

With the exception of dehumidifiers, alternative techniques such as radio frequency (microwave) drying and infra-red radiation, are not used for the drying of British timber. Drying by dehumidifier has the attraction that the same energy is used twice. A mechanism similar to that used in a compression type of refridgerator is employed. Air is compressed by a pump and, in consequence, it becomes hot. It then passes through a relief valve allowing it to expand and as a result it becomes quite cold. Air within the drying chamber is first blown over the coils of chilled air causing much of its moisture to condense and drip away, it then passes

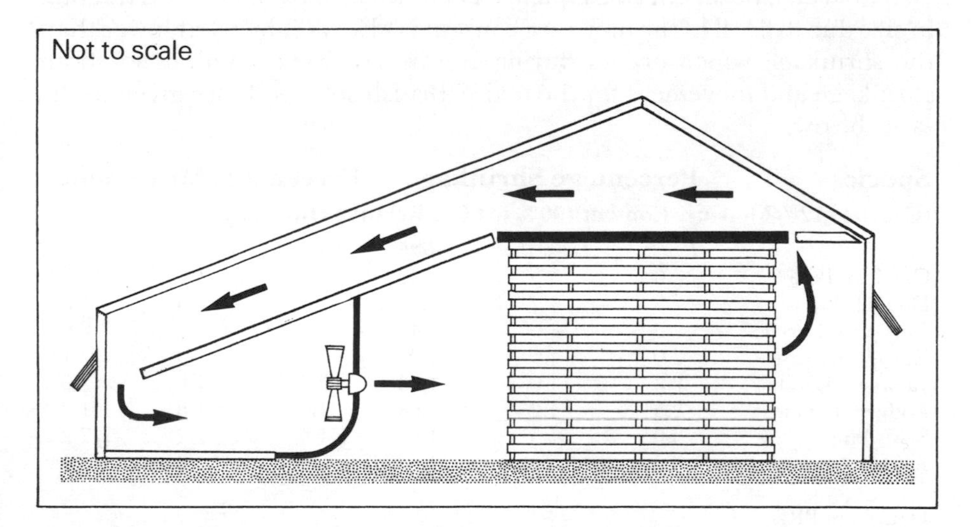

Fig 4.4 Solar drying kiln with fan-assisted air movement; glass or clear plastic and aluminium frame.

over the coils of the hot air, after which it is fanned over the stack of timber as warm dry air.

Solar drying based on the greenhouse principal has been successfully pioneered at the Oxford Forestry Institute (Fig 4.4); but under British conditions the low intensity of solar radiation between October and March means that it is not cost effective during this period. The method is sometimes complemented by dehumidifiers to increase efficiency.

Shrinkage and movement

Shrinkage commences when the wood is dried below its fibre saturation point, and it occurs to such a degree that green timber is normally sawn in sections a few millimetres oversize to allow for it. The extent to which shrinkage takes place depends on the species as well as the way in which it has been converted.

Tangential shrinkage is invariably greater than radial shrinkage (usually roughly double), but no home-grown timber exhibits measurable shrinkage along the grain unless reaction wood (see page 73) be present. Shrinkage can vary from 2 per cent (European spruce; radial) to 5.5 per cent (Corsican pine: tangential).

Once the timber has been dried it remains hygroscopic; its moisture content adjusts according to the relative humidity of the environment. The stable moisture content at which it settles at a given relative humidity is called the equilibrium moisture content. When the ambient relative humidity changes the moisture content also changes and the dimensional change which occurs in consequence is known as movement. A reduction in size due to a fall in the moisture content is always somewhat lower than the shrinkage which occurs during the initial drying. Values for both shrinkage and movement for the major British softwoods are given in the table below:

Species	**Percentage Shrinkage (Green to 12% Moisture Content)**			**Percentage Movement (90% to 60% Relative Humidity)**		
	Radial	*Tangential*	*Ratio*	*Radial*	*Tangential*	*Ratio*
Corsican Pine	3.0	5.5	1.83	0.9	1.6	1.73
Douglas Fir	2.5	4.0	1.6	1.0	1.3	1.3
European Larch	2.8	4.0	1.43	0.8	1.6	2.0
European Spruce	2.0	4.0	2.0	0.7	1.5	2.14
Japanese Larch	2.8	4.0	1.43	0.8	1.6	2.0
Lodgepole Pine	2.5	4.2	1.68	1.0	1.8	1.8
Scots Pine*	3.0	4.5	1.5	1.0	2.2	2.2
Sitka Spruce	3.0	5.0	1.67	0.9	1.3	1.44

Data from PRL

* Scots pine is classified as having a "medium" movement, all the other species listed above have "small" movement.

It is interesting to note that while Scots pine (especially in its imported form as redwood) is traditionally the preferred species for joinery, it exhibits both the greatest movement and the largest differential between radial and tangential movement. However, the worst distortion problem in home-grown softwood is twist resulting from spiral grain, which can be particularly severe in Sitka spruce (the long term solution through the reduction of spiral grain is a task for the forest geneticist).

The specification of kiln dried wood

Difficulties caused by the movement of wood in service make it imperative that the clause in the specification relating to drying takes into consideration the equilibrium moisture content of the environment in which the wood will be used. It is not sufficient to specify ". . . all timber shall be kiln dried. . .", a required moisture content with tolerances should be defined. Additionally, those handling the dried wood should be instructed on the care to be exercised in the transport and storage to ensure that it does not pick up moisture before installation. Problems such as the opening of joints, the appearance of gaps in floorboards, and the withdrawal of tongues in tongued and grooved cladding result from a failure to specify an appropriate level of kiln drying, or from exposure of dried wood to the elements before fixing it in position.

On the other hand little harm can result from using timber at a moisture content two or three per cent below that normally specified; any compression resulting from subsequent swelling is usually accommodated without adverse effects.*

The recommended moisture contents for use in various situations are listed below:

End Use	Required Moisture Content	
Woodwork adjacent to radiators etc	8–9 per cent	Kiln drying essential
Joinery in buildings with continuous central heating, kitchen and lounge furniture	10 per cent	
Bedroom furniture, toys	12 per cent	
Portable buildings, garden furniture	15–16 per cent	Kiln drying desirable
Shelves, boxes and other items in contact with food	19 per cent	
Fencing and estate timbers before preservative treatment	20–25 per cent	Air drying adequate

* Problems occasionally arise when kiln dried wood flooring is laid on an insufficiently dried concrete sub-floor.

5

Defects and the Grading of Sawn Wood

Defects

Defects can reduce the performance of wood in three main ways:

1. They can lower its strength properties substantially and reduce the efficiency of joints.
2. They can make it more difficult to work and finish with machine and hand tools.
3. They can spoil its appearance.

However, a feature which is regarded as a defect for one particular end use may well be considered to be of no consequence, or even an asset for another. For example, if elm boards were to be selected for load-bearing purposes in civil engineering applications, it would be prudent to avoid pieces which display wild grain which could have significantly lower bending strength, but if the same pieces were used for furniture manufacture the wild grain could enhance their aesthetic appeal. Similarly blue (sap) stain in Scots pine could make it unacceptable for those markets — such as internal joinery — where appearance is important, but its suitability for uses where strength (which is almost unaffected by blue stain) is the criterion should not be in question. The early stages of fungal infection can create patterns in hardwoods, typically lighter patches often marked by a black line. Such patterns are valued by wood turners for decorative articles. It is therefore essential to know the prospective end-use before acknowledging a specific feature as a defect.

Rot

Decay resulting from attack by wood destroying fungi is the most serious

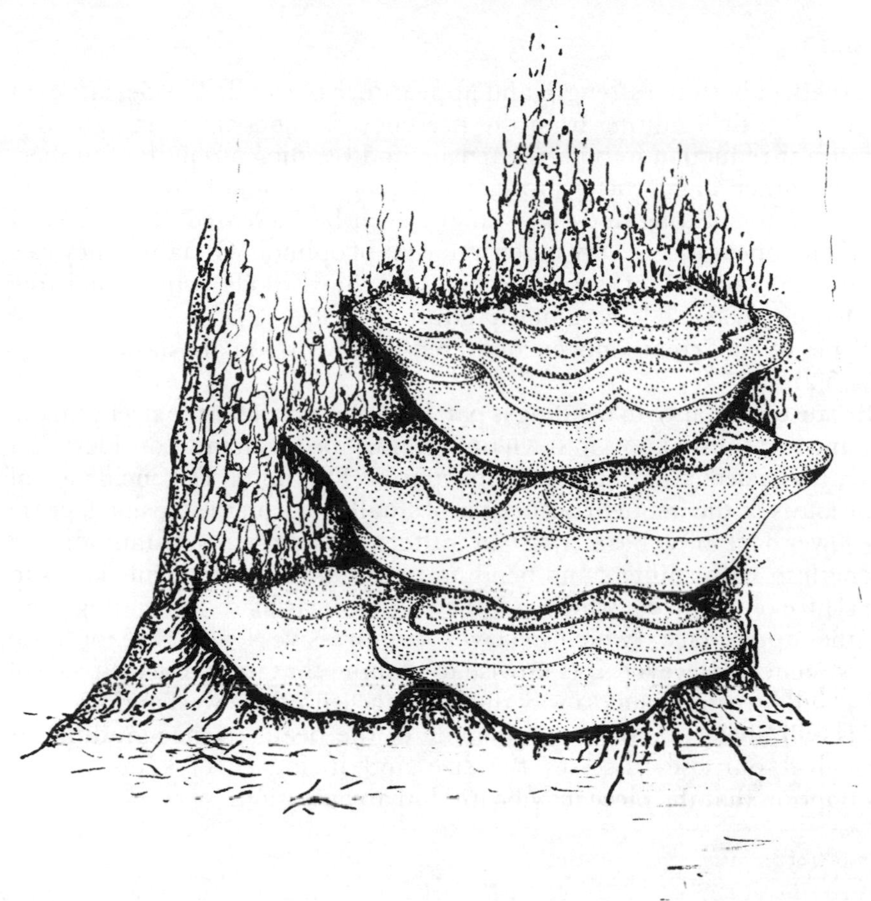

Fig 5.1 Common butt rot — *Heterobasidion annosus* — at the base of a spruce.

defect likely to be encountered in British timber. It can arise in a number of ways. It can be caused by an attack in the living tree by a parasitic fungus such as the common butt-rot (*Heterobasidion annosus*) (Figs 5.1 and 5.4); it can occur as the result of invasion by saphrophytic fungi on undried wood lying in the forest or sawmill; or it can appear on wood in service which has been rewetted (eg by condensation or by plumbing leaks) and subsequently colonised by pests such as the Cellar Fungus (*Coniophora puteana*). The effect of such decay on strength properties can be quite disastrous. Consequently all systems of grading rules for sawn timber make provision for the elimination of, or reduction in, the incidence of decay.

Knots

Knots affect both the strength and appearance of wood. The deviation of the wood grain resulting from the presence of knots can bring about a significant reduction in both the tensile and bending strength of timber. On the other hand the compressive strength is unaffected, while its resistance to splitting and shear might even be increased. The effect of knots on appearance is very much a matter of opinion. Certainly they can reduce the appeal of some ornamental hardwoods while knot-free coniferous wood can look somewhat dull.

Types of knot commonly encountered in sawn wood are shown in Figs 5.2a; 5.2b.

Because resistance to bending is partly a function of tensile strength, the size and location of knots in sawn timber must be taken into consideration when it is selected for structural purposes (Fig 5.2c). The incidence of large knots or the frequent occurance of small knots on the tension face (ie the lower face of a beam, joist or rafter) will reduce substantially its strength in both stiffness and bending to ultimate failure, while knots in the centre of the piece (which is subjected to shear loads in bending) and on the upper face (which receives compressive loads) are of little consequence. However, it is unwise to assume that in practice a piece of wood will be used in such a way that the face with the fewest and smallest knots will always be on the underside of the beam. Hence grading for structural purposes assesses the size and incidence of knots on the assumption that the piece may be used in any position.

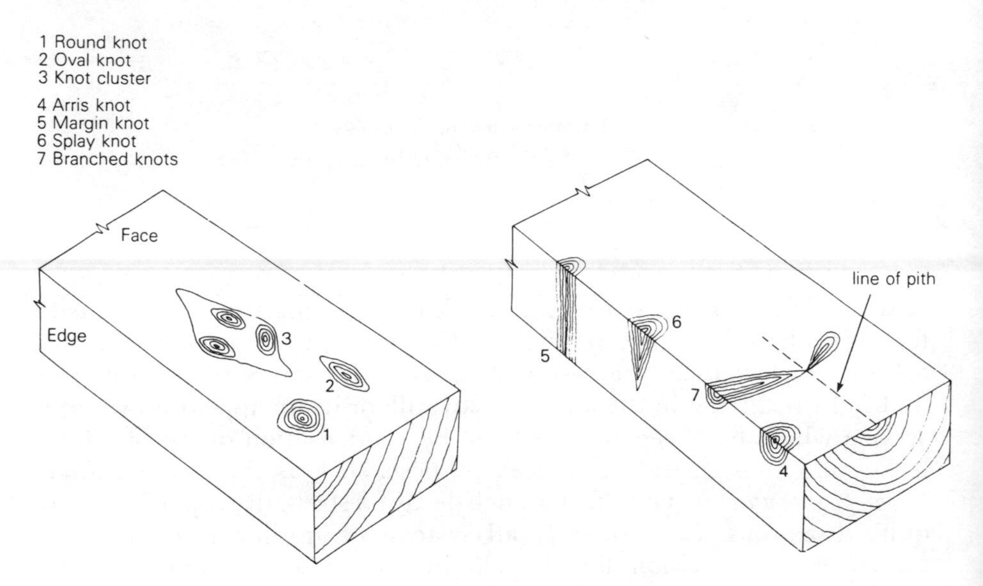

Fig 5.2a Knot types.

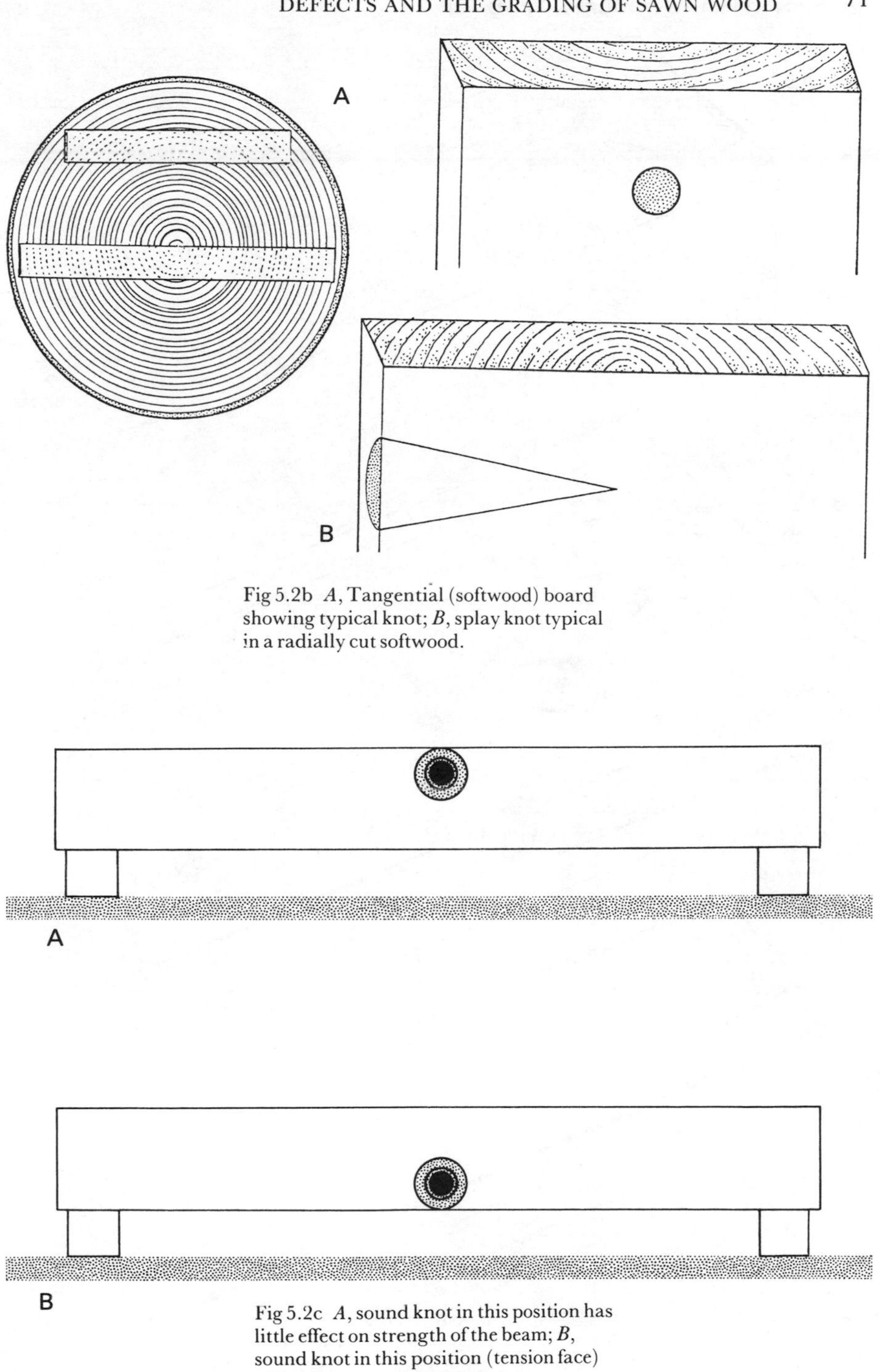

Fig 5.2b *A*, Tangential (softwood) board showing typical knot; *B*, splay knot typical in a radially cut softwood.

Fig 5.2c *A*, sound knot in this position has little effect on strength of the beam; *B*, sound knot in this position (tension face) will reduce strength of the beam.

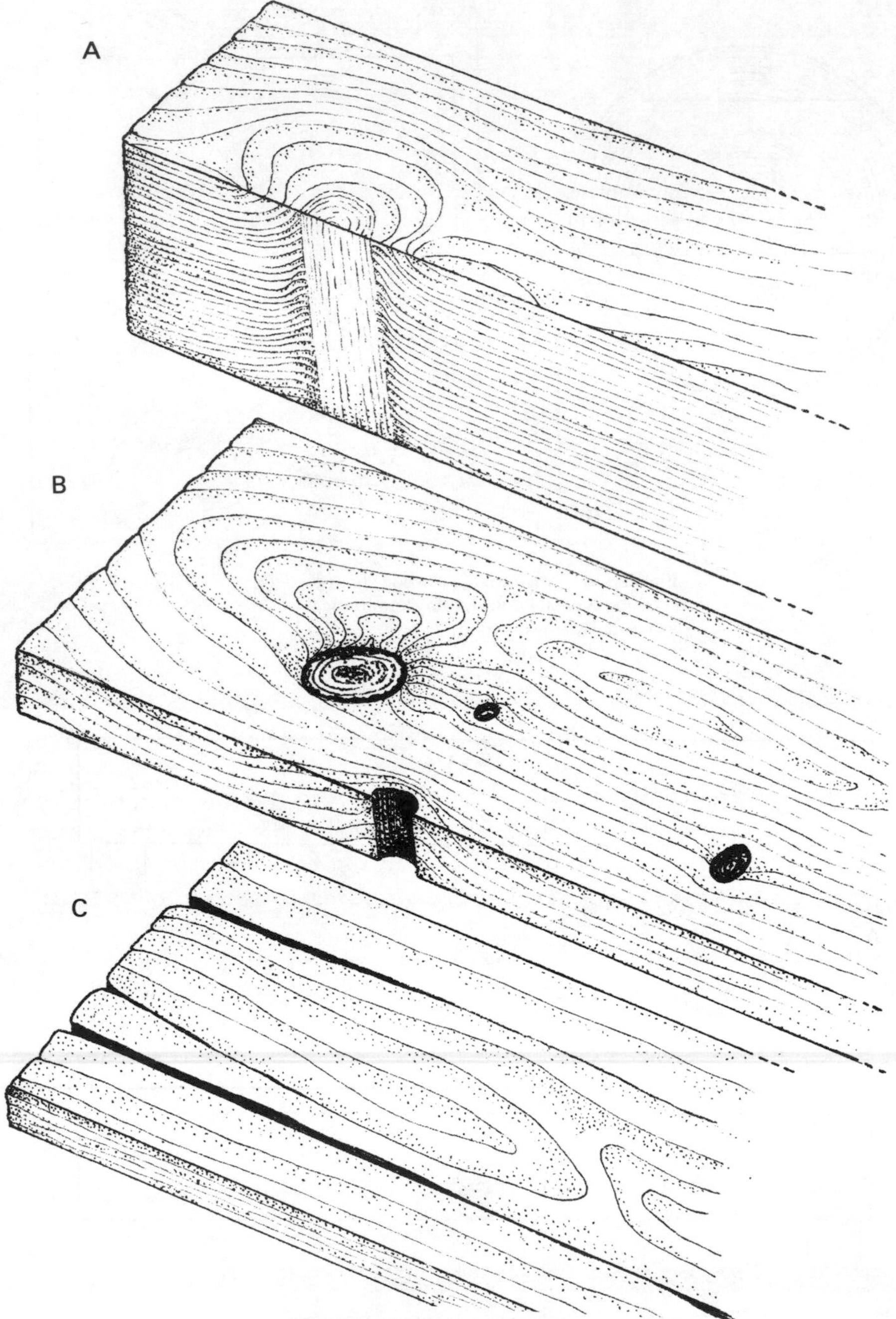

Fig 5.3 *A*, live knot in softwood;
B, dead knots and knot hole in softwood;
C, longitudinal grain fissures, splits or checks in softwood.

In those situations where the wood is to be used for structures where appearance is of no importance, the type of the knot does not affect the grading. Thus live knots and dead knots or knot holes are all regarded as having the same weakening effect on the tensile and bending strength of the wood. Live knots are those originating from branches which are alive at the time of felling: dead knots are formed by the continued radial growth of the stem over branches which are already dead; dead knots are frequently loose and are often encased by a complete circle of bark (Fig 5.3). When wood is used for non-load bearing purposes care should be taken to exclude dead knots from points likely to receive nails or screws, or which are to be morticed.

Knots also affect the working properties of wood. In some species such as spruce the knots (which are hard) are situated in a matrix of relatively soft wood tissue, and this can result in damage to the cutters of fast rotating machine tools. This type of damage can be reduced considerably by a technique known as jointing in which a second back bevel is ground on to the cutting edge. Irregular grain which invariably occurs above and below knots can cause tearing of the wood tissue during machining.

Slope of grain

Wood is seriously weakened if the natural orientation of the fibre deviates greatly from the long axis of the piece. This defect can have more than one cause.

It can result from spiral grain which is a natural phenomenon in both coniferous and broadleaved timbers. In some species such as sweet chestnut it is manifest as a spiral pattern on the bark. It is an inherited character, and as already noted its reduction in the growing crop is largely a matter for the geneticist. When a log with spiral grain is converted the timber sawn from it inevitably has a steep slope of grain. In other species such as elm the grain deviates without showing any fixed or predictable pattern (wild grain). When this occurs it is best avoided for load-bearing purposes.

Sloping grain can also appear in relatively straight-grained species such as pine or sycamore as a result of faulty sawmilling where the saw cuts have been allowed to deviate appreciably from the line of the grain, or from the conversion of crooked logs.

Reaction wood

Reaction wood (Fig 5.4) is tissue which displays inferior strength coupled with excessive shrinkage, especially in the longitudinal plane. It is generally agreed that it is formed in response to a gravitational stimulus; it is frequently present when the growing tree deviates from the vertical, and

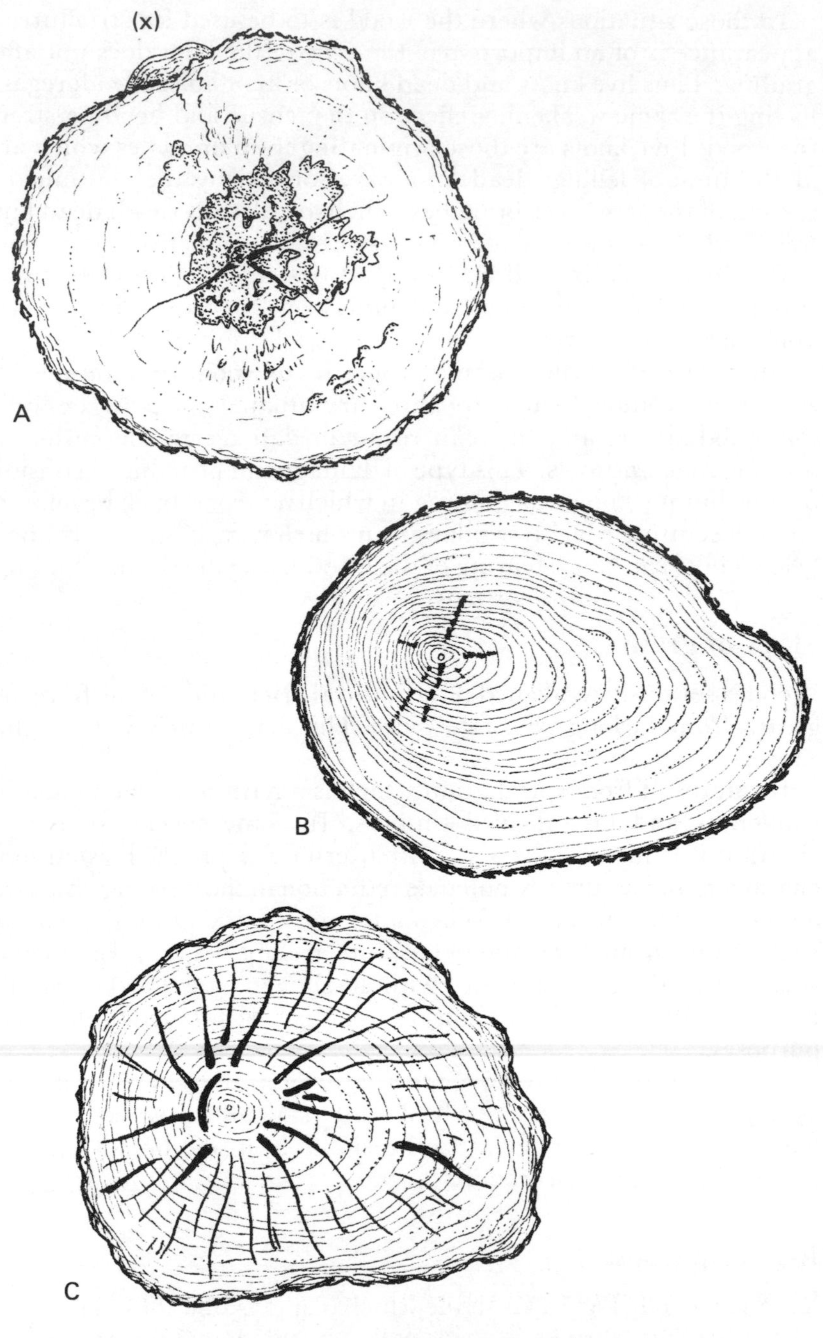

Fig 5.4 *A*, butt rot — note fruiting body at (x);
B, compression wood — Scots pine;
C, fissures and tension wood — beech.

it is invariably present in branchwood.

In conifers it is termed compression wood being found on the lower side (or compression face) of branches where it is associated with a somewhat larger ring width; while in broadleaf trees it is known as tension wood occurring conversely on the upper or tension side of branches where it is also associated with wide rings.

Compression wood is characterised by rounded thick-walled short tracheids with intercellular spaces, but it is identifiable to the naked eye by an increase in opacity or as a darker colour than the adjacent tissue. It is appreciably weaker in most strength properties, and timber displaying excessive development of compression wood is best avoided for structural purposes. However, even the most vertical trees contain some compression wood; for example, it appears in the tissue immediately beneath knots. Where it is present to an undesirable extent, it can be recognised and "rogued out" after drying because its excessive longitudinal shrinkage is likely to cause an unacceptable degree of distortion; wood in this condition will almost certainly be rejected in machine grading.

Tension wood consists largely of fibres with a gelantinous appearance, with little or no lignification. It exhibits abnormally large shrinkage, and it is particularly weak in resistance to compression. Boards containing a high proportion of tension wood can similarly be "rogued out" after drying, when the defect is often manifest as wash-boarding (i.e. a ridged surface running across the grain).

Wane

Wane is defined as the original rounded surface of the log (with or without the bark) remaining on sawn wood. Where it is small, and the appearance of the wood is unimportant, it can be tolerated for several end uses; but where it exceeds more than a few millimetres in depth or width, it can result in a significant weakening of the piece. Sawlogs can of course be converted in such a way that no wane is produced, but the tolerances for this defect in most specifications are such that it is usually uneconomic to do this.

Stains

Stains can result from colonisation by otherwise harmless fungi (see chapter 8), or from chemical reactions between metals and extractives; eg between iron in metal fastenings and tannins in species such as oak. While such stains reduce the value of the wood for ornamental purposes they do not affect the inherent strength to any noticeable extent.

Rate of growth

Coniferous timber displaying widely spaced annual growth rings has often been regarded as having inferior strength properties, and there has been a tendency to use the number of rings per 25 mm (or per inch) of radius as an index to assess the strength properties of softwoods. This is a fallacy which has persisted despite the substantial volume of evidence to the contrary. It is not uncommon to find moderately wide-ringed timber with good strength properties, especially in Douglas fir and the three-needled pines. Nevertheless, extremely fast grown or slow grown wood is best avoided for load-bearing purposes in both softwoods and ring-porous hardwoods.

Although any statistical correlation between ring-width and strength, or ring-width and density is usually quite low, some softwood grading rules place a lower limit on the number of rings per 25 mm of radius. This is done for two reasons; where the timber is to be used for structural purposes specifying a minimum number of rings has the effect of reducing the proportion of juvenile wood, if present (see below). Similarly, in some

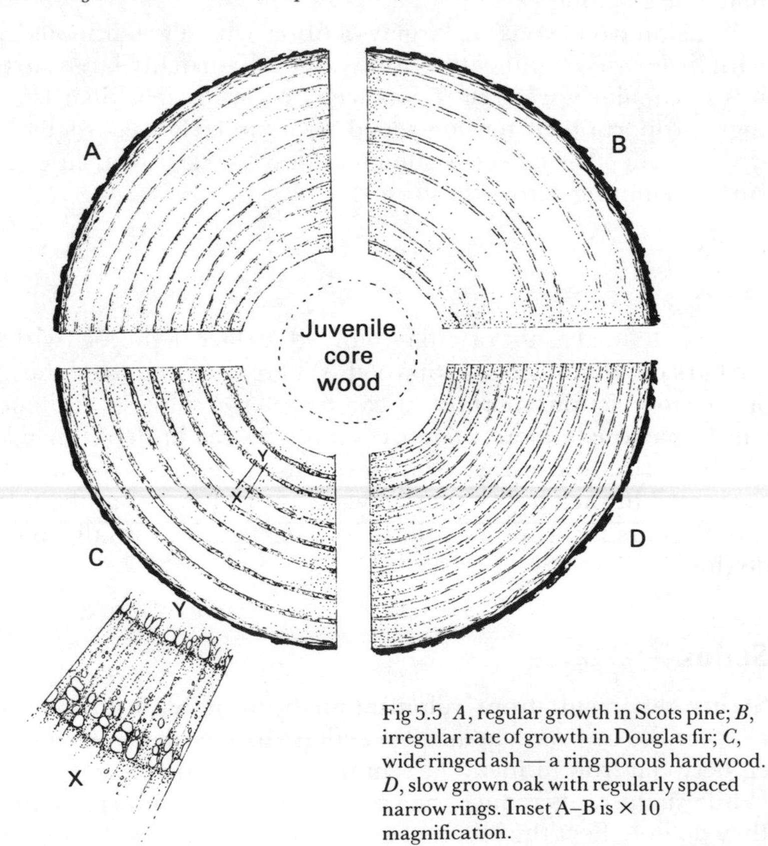

Fig 5.5 *A*, regular growth in Scots pine; *B*, irregular rate of growth in Douglas fir; *C*, wide ringed ash — a ring porous hardwood. *D*, slow grown oak with regularly spaced narrow rings. Inset A–B is × 10 magnification.

of the lower density species such as spruce, there is a relationship between the quality of the finish after machining and the width of the rings. Wide-ringed wood can give a woolly finish, and for this reason the current British Standard for joinery requires a minimum of four rings per 25 mm of radius (six for external joinery).

In ring porous hardwoods the opposite situation applies because narrow-ringed wood resulting from slow radial growth displays a high proportion of pores (vessels) at the expense of the dense fibrous tissue from which the timber derives its strength.

Where ash is required for tool handles, sports goods, or other purposes where shock loads are likely to be experienced, it is imperative to avoid narrow-ringed wood. The same principle applies in the use of oak or ash for ladder rungs.

Juvenile wood

The wood laid down by a tree before it reaches the cone-bearing — or flowering — stage, is somewhat different in character and generally inferior in strength to the wood produced by the older tree. This tissue, which is confined to the inner 10–20 growth rings, is termed juvenile wood.

In high value hardwoods it is possible to remove the juvenile wood and the knotty core during sawmilling by a technique called "boxing out the heart", but in softwood and lower grade hardwoods it is not economic to do this.

Splits and checks

Fissures along the grain arise mainly as a result of shrinkage during drying, (Fig 5.3) but they can also occur in consequence of the uneven relief of stresses in the growing trees in felling or during sawmilling. They can affect the strength performance, especially in bending, shear and nail-holding; and their incidence is often limited in grading rules.

Grading rules

There are currently no grading rules specifically for sawn British timber. Grading rules for softwoods were published in 1964 as British Standard 3819, but no metric version was drafted, and this Standard was withdrawn following the publication of British Standard 4978 "Timber grades for structural use" which is applicable to all commonly available softwoods irrespective of their country of origin. It provides for the selection of two grades namely GS (General Structural) and SS (Special Structural).

Those involved in the formulation of grading rules for coniferous timber grown in Britain avoided imitating the Scandinavian system (also used in Poland and Eastern Canada) with its contradictory nomenclature*, instead they attempted to devise a system of grading by end-use requirements. Thus during its short life BS3819 prescribed four grades, one suitable for joinery, one for structural work, one for high grade packaging, and one for general purpose packaging.

However, as already noted, the introduction of BS4978, which incorporates the strength correlated *knot area ratio* method of stress-grading, rendered them obsolete. The present position, therefore, is that those wishing to market joinery quality wood should select in accordance with the requirements of British Standard 1186 "The quality of timber and workmanship in joinery" Pt. 1 (1986) (i.e. this Standard specifies a rate of growth of no fewer than 4 annual rings** per 25 mm of radius, and places restrictions on the incidence of knots, especially dead knots), and those wishing to supply structural timber should either grade the wood visually using the knot area ratio method, or arrange for it to be stress-graded by machine.

The knot area ratio (KAR) concept was introduced during the 1960s; prior to its conception, grading systems for softwoods had shown embarrassingly low correlations between grade and strength. It is based on the principle — already noted above — that knots on the tension face of a structural member cause a far greater reduction in bending strength than knots elsewhere; hence, in this system the knots arising on or near the edge of a beam rafter or joist are more severely restricted than knots near the centre. As the grader does not know which way round the timber will be placed, the restriction is applied to both edges. BS4978 also requires the number of growth rings to be a minimum of 2½ per 25 mm of radius to keep the proportion of juvenile wood low; it also restricts the incidence of wane, and limits the slope of grain.***

The grading of hardwoods is much more complex. The former British Standard 4047, which was seldom enforced, made provision for a "defects system" and a "cutting system", and on to these additional grades were superimposed namely "pin-knot no defect". The defects system operated

* Scandinavian timber is normally imported in three grades based on the incidence of defects and knots. The highest grade is called "unsorted" and is commonly specified for joinery; the second grade known as "fifths" has been widely used for structural timber; and the third and lowest grade is called "sixths" or "Uttskott" and is in demand for packaging and materials handling. Russian grades are similar, but are named "unsorted", "fourths" and "fifths" respectively. It has to be emphasised that these grades are based solely on appearance, and are not necessarily related to strength.

** Six rings per 25 mm for external joinery.

*** British Standard 4978 : 1988 "Specification for softwood grades for structural use restricts the incidence of wane to one third of the thickness, and one third of the depth for the full length, together with one half of the width or depth in any 300 mm length. The slope of grain may not exceed one in six; and the average annual ring width may not be greater than 10 mm (= 2½ per 25 mm of radius).

in a similar way to softwood grading, and it could show some correlation with the strength of the piece. However, the cutting system was much preferred by those consumers such as furniture manufacturers who use hardwoods for their ornamental effect. The definition of grades under the cutting system is complicated, and its operation is not easy for the newcomer to comprehend, but basically, the grades are related to the proportion of clear timber which can be cut from the piece.

In practice the formal grading of British hardwoods is not widely used. Instead the timber merchant selects to meet the requirements of individual customers.

Mechanical stress-grading

It has been established for many years that there is a firm statistical relationship between the ultimate bending strength of wood (modulus of rupture) and its stiffness (Young's modulus). Hence as it is possible to assess the stiffness non-destructively by either noting the deflection given by applying a known load, or by recording the load needed to cause a given deflection, a good prediction of the ultimate strength can be made without breaking the specimen under test. This principle is the basis of stress-grading by machine.

Stress-grading machines have been operating in Britain for two decades. In most types, the sawn wood is fed through the machine by powered rollers; in some systems a constant bending load is applied centrally over a fixed span and the resultant deflection is measured automatically at predetermined intervals, and the data are fed continuously into the machine's computer; in other types the load required to produce a standard deflection over a given span is recorded repeatedly and the information is assembled in the machine's computer. In both systems the computers are linked to dye applicators which colour the wood at intervals to indicate the strength category at that point.

There are a number of advantages in grading by machine:

1. With outputs in the region of 4,000 linear metres per hour it is much quicker than visual grading.
2. It entails greater precision than visual grading in that it virtually guarantees the strength of a piece.
3. It can be set to the lower limits of the strength categories defined in British Standard 5268 "Structural Use of Timber" thus giving optimum yields.
4. It gives higher yields than are obtainable by visual grading because it transcends features such as edge knots and ring width which could necessitate the rejection of a piece under the knot area ratio method. In some instances thirteen fold increases have been recorded in the yield of timber meeting a required grade.

5. It identifies "weak spots" in such a way as to suggest reconversion. For example if a weak point were to appear in the centre of a five metre joist it might be possible to convert it into two, 2.5 metre joists of a higher grade.

The implications of machine stress grading in both traditional housing and timber frame structures are considered in Chapter 7.

International standards for stress grading and finger-jointing of structural coniferous sawn timber

The Timber Committee of the Economic Commission for Europe has been working since the early 1970s on the standardisation of rules for stress grading and finger-jointing of structural coniferous sawnwood. The Timber Committee has now terminated its work in this field which will be continued by such bodies as the International Organisation for Standardisation (1S0). Its recommended standards have been published as an internationally agreed reference point for the work being carried out by other bodies, as "ECE recommended standards for stress grading and finger-jointing of structural coniferous sawnwood" ECE/TIM/45, United Nations, Geneva (SE.89–45268, April 1989).

The introduction reads:

> This recommended standard lays down stress grading rules for both visual and mechanical grading of coniferous sawn timbers. The grading rules are designed primarily, but not exclusively, to separate the major European construction softwoods into three distinct grades, for which well defined characteristic strength and elasticity values can be assigned.
>
> Where North American timbers are graded to the National Grading Rules for Dimension Lumber, they are also acceptable for structural use.
>
> This recommended standard does not cover scaffolding boards or glulam.
>
> Where construction softwood includes a variety of species, it may be advantageous for strength classes to be considered by the specifier as an alternative to specifying species and grades. National standards can readily relate strength classes to the appropriate stress grades specified.
>
> It should be noted that systems for defining structural timbers based on strength classes have been introduced into national and international codes (e.g. CIB-W18). Most strength classes, where they are specified, may be satisfied by various combinations of species and national and international visual grades, or by special selection by grading machine.

6
Fencing

Types of Fencing

Fencing is a major market for British timber. It absorbs a wide range of species sizes and qualities of both hardwoods and softwoods — round, half-round, sawn and cleft. Given adequate preservative treatment no species is wholly unsuitable.

The four main types of fence are:

- Post and rail fences, and post and wire fences for stock control on farms (Figs 6.1 and 6.2).
- Forest fences, mostly post and wire, to exclude stock, rabbits and deer from woodlands (Fig 6.3).
- Roadside and motorway fencing mainly for the exclusion of stock from major roads; post and rail, and post and wire are most commonly used, although some close-boarded fences may be erected in special situations where screening and noise reduction is required.
- Domestic garden and urban fencing where panels and close-boarded fencing predominate to provide privacy and ornamental effect (Fig 6.4).

In addition, there are two familiar sorts of fence in limited supply, namely cleft chestnut paling (Fig 6.5) which is used mostly as a portable guard fence, and hurdles woven by craftsmen from hazel coppice which is used mainly for sheep folding and for garden screens (Fig 6.6).

Sizes of timber required

Post and rail fences, and post and wire fences The wooden components in post and wire, or post and rail fences range in size

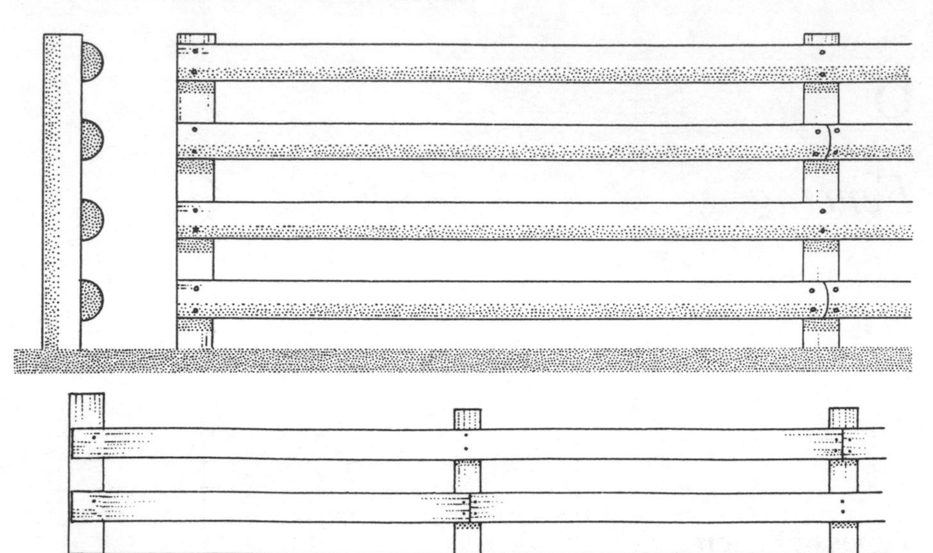

Fig 6.1 Half round post and rail stock fence.

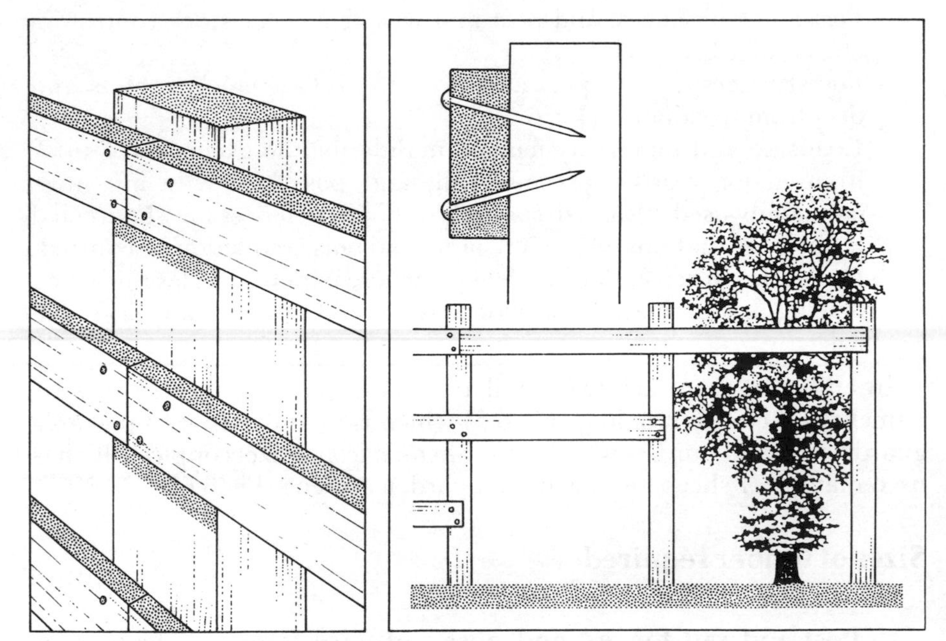

Fig 6.2 Post and rail fence in sawn softwood with nailing detail.

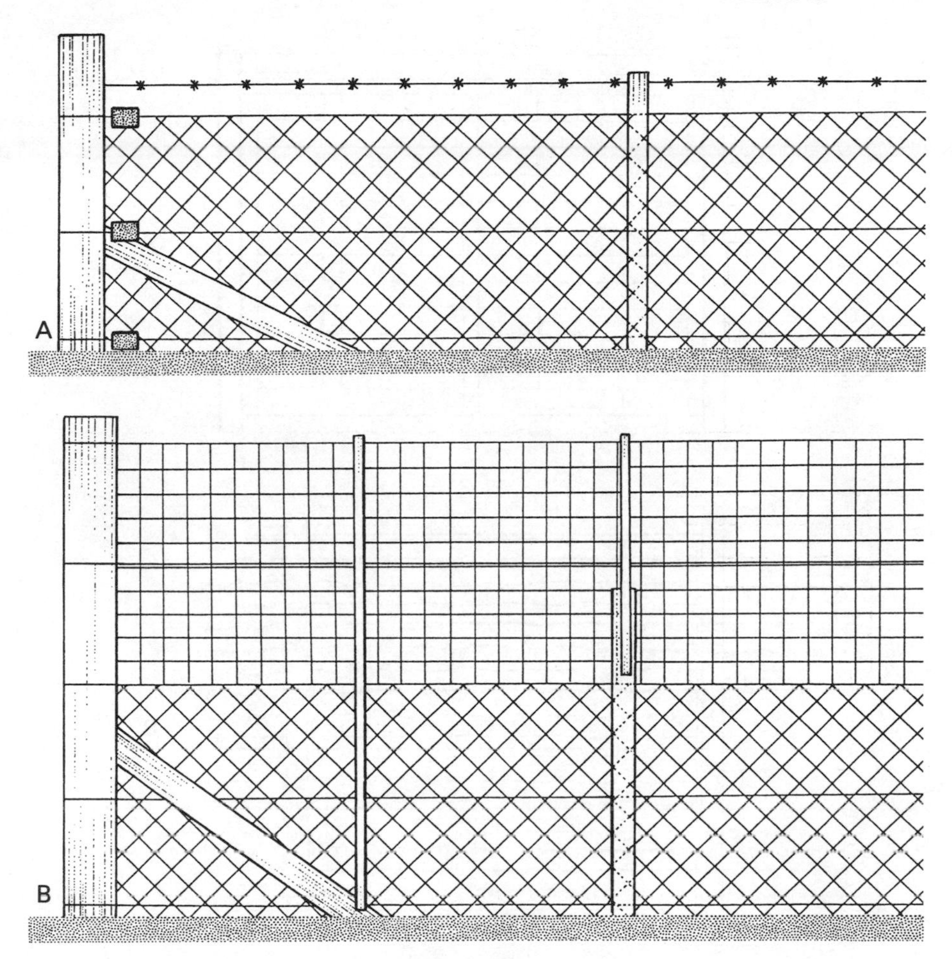

Fig 6.3 *A*, stock and rabbit fence between field and woodland; *B*, stock, deer and rabbit fence between pasture and woodland.

from light round stakes around 50 mm top diameter and 1.5 m long used for forest rabbit fencing to the stout strainers used in deer fencing which may have a minimum top diameter of 175 mm and a length of 2.75 m (a Scottish specification). Some specifications allow the posts or stakes (stobs) to be round, half-round, quartered or sawn provided that the cross section does not fall below a prescribed minimum area; others, such as those for post and rail fences, permit the use of rectangular or sawn square posts with narrow tolerances on the dimensions because they must be able to accommodate the

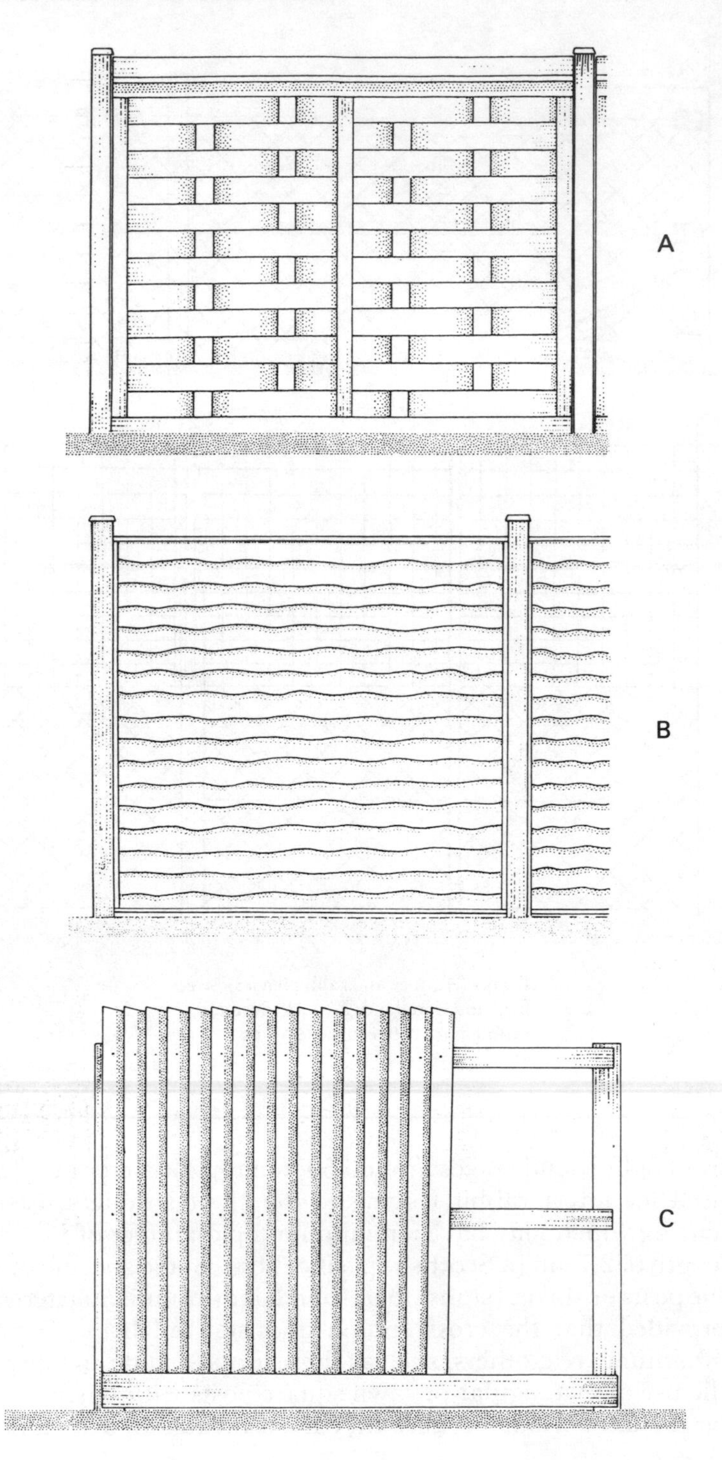
A
B
C

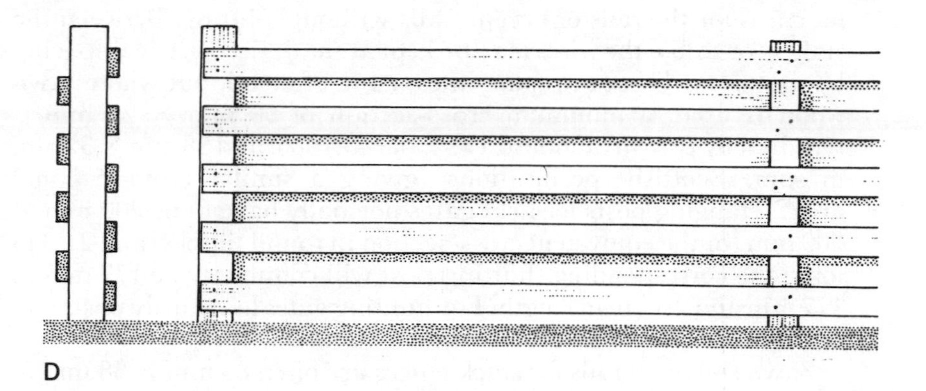

Fig 6.4 *A*, interwoven fence,
B, overlapped fence;
C, vertical feather-edged boarding fence;
D, ranch type fence;
E, sawn oak pale fence erected 1959.
Photographed 1989.

mortices for the rails or accept nails without splitting. Between the stakes (or stobs) the wires may be kept at the desired vertical spacing by droppers. These are sometimes made of metal, but where sawn wood is used, a minimum cross-section of 50 mm × 25 mm is required by British Standard 1722, part 3:1986, and 38 mm × 32 mm in some Scottish specifications (giving a similar cross-sectional area). Hanging posts for field gates normally have to be 200 mm × 200 mm (or the equivalent cross-section in round timber) and 2.43 m long; the corresponding shutting post will commonly be 177 mm × 177 mm and 2.13 m in length. For hunting gates less sturdy posts will suffice.

Sawn softwood rails for stock fences are often 89 mm × 38 mm in cross-section (87 mm × 38 mm in BS 1722, part 7:1986 — the current and metric versions of the Standard) and 2.75 or 2.9 m long depending on whether they are to be butt jointed and nailed on to the posts, or to be morticed, which requires an overlap. Motorway fence rails are 100 mm × 30 mm in softwood, or 90 mm × 40 mm in hardwood. If fencing components are cleft, ash is the species used most frequently, combining high strength with relative lightness. Oak and sweet chestnut are also used for cleft fencing; local availability of suitable material and tradition determines which species is used. Half-round softwood rails nailed to half-round softwood posts give a good durable post and rail fence if both components are treated with a preservative under pressure.

Large quantities of round timber in the form of short logs known as "fencing bars" are used by the manufacturers of garden and urban fences. The minimum size normally in demand is 1.85 m length by 140 mm top diameter. Most softwoods are used but larch is the preferred timber. The "bars" are cut into thin boards and the other components from which woven and lapped panels are made. In these types of fence a ready made panel supports itself between the posts (Fig 6.4). It has the advantages of ease and speed of erection, as well as being less expensive than post and rail fencing clad with vertical boards of cleft or sawn oak or sawn softwood. Cleft boards are often difficult to obtain, and they have been largely superceded by feather-edged sawn wood which is typically 100 mm wide by 13 mm tapering to 6 mm (BS 1722, part 5: 1986).

Other types of urban fence include vertical spaced pale fences; horizontal board fences made of 150 mm or wider softwood boards 25 mm thick nailed to the wooden posts on one or two sides (Fig 6.4).

With the exception of rails which have to be straight-grained and free from large knots, much fencing timber can include features which would make it unacceptable for joinery or construction. As most fencing timber is of short length, compared to wood used in building, defects can be cut out with less waste.

British Standard 1722 gives guidance on the assessment of defects and the extent to which they are permitted. The list includes knots, slope of grain, rate of radial growth, checks, splits, resin and bark pockets, rot, insect damage stain, distortion and boxed heart.* The dimensions of timber components assume a moisture content of 28 per cent and guidelines are given on the tolerances which are allowed if the timber components are below this level at the time of use. An allowance (reduction) on the specified size of one per cent for every five per cent in moisture content below 28 per cent is permitted, for example, in BS 1722, part 7: 1986. Although British Standard 1722 gives comprehensive information on the dimensions and qualities needed for the various components, it is essential in practice to ascertain what individual customers expect and are prepared to accept. A major confounding factor is the continued use of both imperial and metric units. Some purchasers will accept the conventional rounded-down metric equivalents, but others will not.

Cleft chestnut fencing Cleft sweet chestnut fencing is made from half-round or roughly triangular pales held together by two or more lines of twisted wire (Fig 6.5). It is widely used in temporary anti-intruder fences in situations such as building sites or road

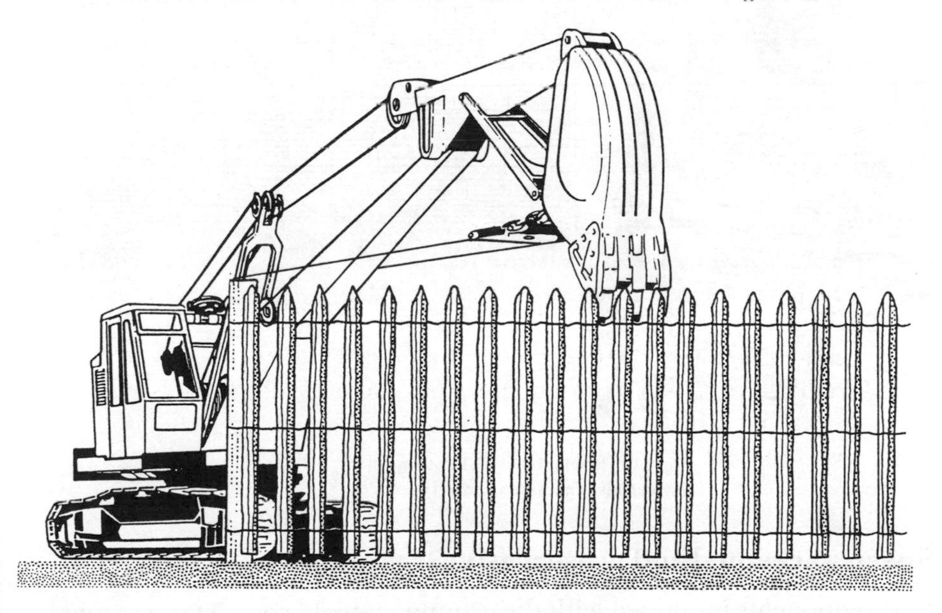

Fig 6.5 Cleft chestnut fencing

* Boxed heart refers to the inclusion of the central section of the tree in a sawn section. It contains the pith, much juvenile wood, and is likely to be knotty.

works. It is available in heights up to 1.8 m, with gaps of 50, 75, or 100 mm between pales. The pales must have a minimum girth of 100 mm to comply with the British Standard. The chestnut from which they are cleft is coppice-grown, and, together with hop poles, it is the main use for the extensive areas of managed sweet chestnut coppice in south eastern England — notably in Kent.

Hazel hurdles The production of garden hazel hurdles (six feet long by three, four, five or six feet high) was in the region of 13,000 dozen per annum in the mid 1950s. Today, only a few craftsmen remain, producing them for sheep-folding, and for garden screens and fences, mainly in central southern England. The hazel coppice for hurdle making is cut when it is nine or ten years old (Fig 6.6), and woven while it is still "green", if it were allowed to dry it would not be pliable enough.

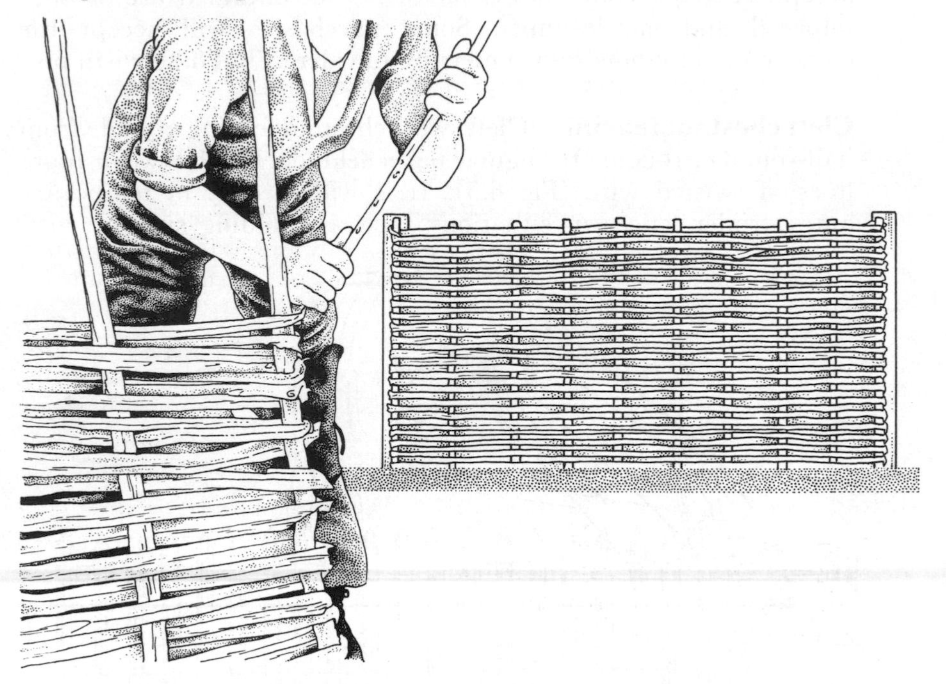

Fig 6.6 Hazel hurdle maker using cleft hazel rods from nine year old coppice.

Species accepted and preservative treatment

For components in contact with the ground, namely posts, stakes (stobs), strainers and struts, only the heartwood of oak, sweet chestnut and yew can be relied upon to give a service life exceeding 25 years in all situations.* In standard tests the larches have shown a wide range of durability and this is reflected in practical experience. As a result the heartwood of larch has been variously described as "naturally durable" or "moderately durable" (life expectancy in contact with the ground of 10–15 years).

This variation in the performance of larch fencing material in contact with the ground can be explained by the fact that the outer heartwood is more resistant to decay than the inner heartwood which contains the central core of juvenile wood. Trees which get off to a good start and make rapid early growth have a substantially wider core of juvenile wood than those which grow slowly in their earlier years.

Extensive testing has shown that, comparing like with like rates of early growth, there is little difference in the natural durabilities of the heartwood of European and Japanese larches and their hybrid (Dunkeld larch). The high reputation of European larch, at one time considered to be much superior to Japanese larch, is probably due to the fact that much of it grew slowly in its early years, in times when conifers generally were planted with less site preparation than today, and at relatively close 0.9 m spacing. Many of these plantations were also affected by larch canker which significantly reduced their rate of growth (Fig 6.7).

For any given site, decay tends to spread into wood in contact with the ground at a constant rate, hence the shape of the cross-section has an important influence on the service life. For example, a square post 106 mm × 106 mm can be expected to last longer than a rectangular post of the same species 150 mm × 75 mm which has the same cross-section area. Of course in designing a fence other factors have to be considered such as the anticipated direction of the main load, the need to cut mortices, and the provision of an adequate nailing surface. Consequently the final shape of the fence posts may well be a compromise.

Preservative treatment is covered in Chapter 8. But it is pertinent to stress here that as round timber is more easily impregnated — on account of its outer layer of sapwood — it gives a better performance than sawn timber for posts and other items in contact with the ground. On the other hand the use of round posts may be impracticable where a good flat nailing surface is needed, in which case the choice is limited to sawn or half-round posts.

British standards

British Standard 1722 specifies in detail the requirements of all the components of fences and the method of erection. It is published in

* The performance of fence posts is also dependent upon the environment. Posts driven into infertile or water-logged soils in areas of high rainfall last much longer than posts in arable soils in low rainfall localities. Untreated spruce in peat in high rainfall areas has lasted more than 20 years.

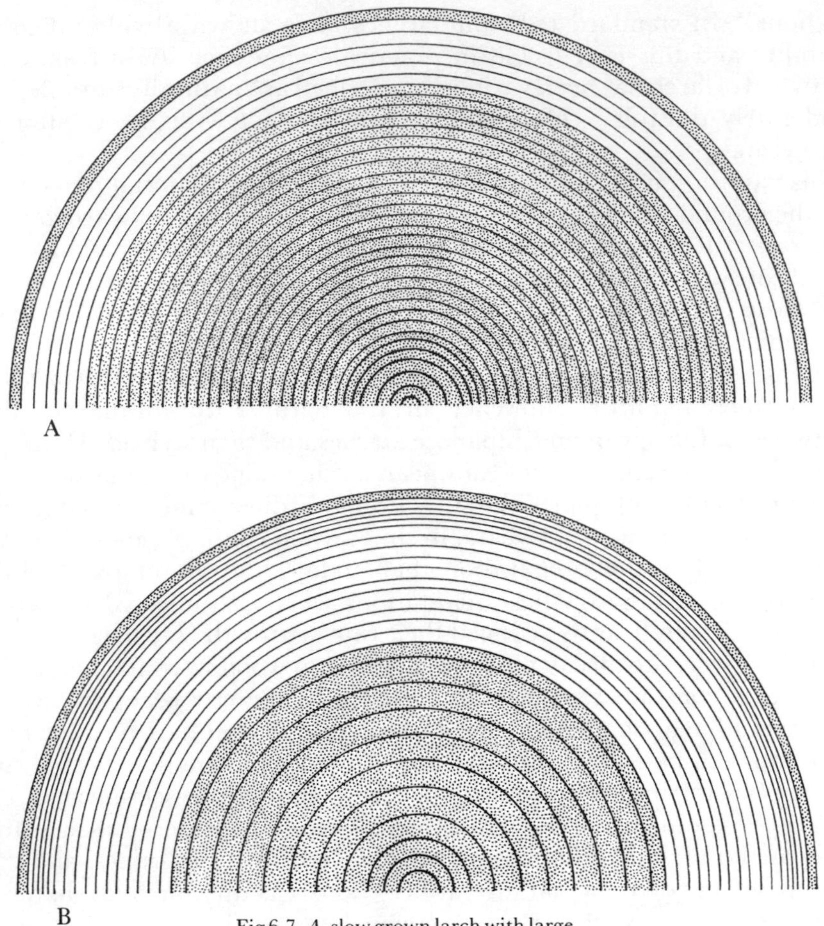

Fig 6.7 *A*, slow grown larch with large proportion of heartwood. Planted at close spacing: thinned; *B*, fast grown larch with small proportion of heartwood. Planted at wide spacing: unthinned.

thirteen parts; those which refer to wooden fences or fences with essential timber components are:

Part 1. Chain Link Fences.
Part 2. Woven Wire Fences.
Part 3. Strained Wire Fences.
Part 4. Cleft Chestnut Pale Fences.
Part 5. Close Boarded Fences.
Part 6. Wood Palisade Fences.
Part 7. Wood Post and Rail Fences.
Part 11. Woven Wood Fences.

This Standard also contains references to relevent British Standards and Codes of Practice for the preservative treatment of timber for use in fencing.

Electric fences

The use of electric fences is increasing. Permanent fencing may be all-electric, or a combination of conventional fencing with one or two electrified wires.

While the requirements of the wooden components are much the same as for conventional strained wire fencing, round posts have an advantage over sawn posts because their shape enables the electrified wire to be held clear of them, especially on corners and bends.

7

Markets for Sawn Wood other than Fencing

During the war years 1939–45 an enormous strain was put on British woodlands* by the urgent demand for both hardwoods and softwoods. Excluding derivative products such as charcoal, from 1939 to 1945 the production of home-grown hardwoods averaged 2.78 million cubic metres a year (roundwood equivalent under bark measure); this is four and a half times greater than the average home production between 1934 and 1938. The corresponding figure for net hardwood imports was 0.8 million cubic metres, giving an apparent consumption of 3.58 million cubic metres (ie home production plus imports, less exports).

By 1950 the production of hardwoods from British woodlands had fallen to 1.65 million cubic metres, and imports had risen to 2.8 million cubic metres; in 1955 the figures were 1.45 and 2.3 million cubic metres respectively.

Excluding derivatives, production of British coniferous timber (which averaged only 0.85 million cubic metres a year between 1934 and 1938) fell from a 1939–45 average of 3.8 million cubic metres a year (roundwood equivalent under bark measure) to 0.87 million cubic metres in 1950 and 1.13 million cubic metres in 1955.

During this time British hardwoods accounted for a higher proportion of the total hardwood consumption than did British softwoods of the total softwood consumption (Fig 7.1).

In addition there were considerable imports of wood products made from hardwoods and softwoods called "derivatives of wood" or "derived wood products". These included pulp, paper and panel products (especially plywood and hardboard), as well as a whole range of items

* Imports, fellings, and the conversion and sale of timber were controlled during and immediately after the war years. Softwood consumer licencing ended in 1953.

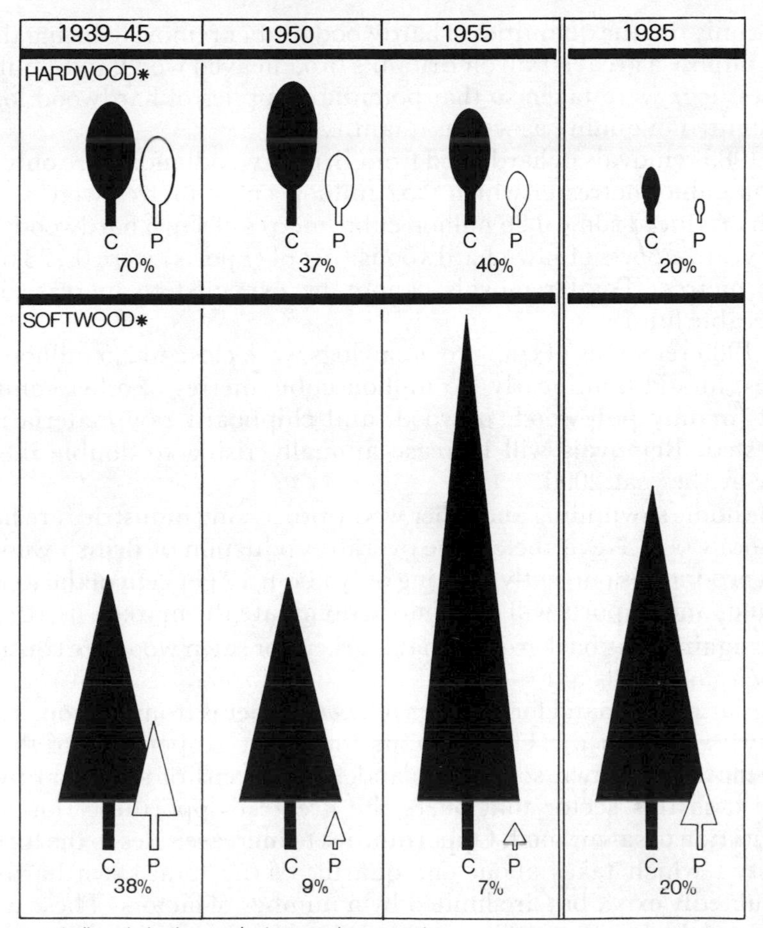

Fig 7.1 United Kingdom production (P) of hardwoods and of softwoods as a percentage of total consumption (C) of hardwoods and of softwoods.

such as furniture, matches, charcoal and veneers. The trend is indicated in the table below:

Annual net imports of derivative wood products — millions of cubic metres.
(Roundwood equivalent under bark measure)

1934–38	1939–45	1950	1955
12	4.7	7.6	13.3

Source: Great Britain Statements to the Empire and Commonwealth Forestry conferences of 1947, 1952 and 1957.

Not only did the quantities of hardwood timber removed during the war years impose a great strain on Britain's broadleaved woodlands, but often the best logs were taken so that potential supplies of hardwood logs are now limited in quality as well as quantity.

In 1985 removals of hardwood from British woodlands were only 0.835 million cubic metres of which 0.52 million cubic metres were sawlogs, which produced some 0.26 million cubic metres of sawn hardwood; in the same year imports of sawn hardwoods (net of exports) were 0.728 million cubic metres. Total removals cannot be expected to increase in the foreseeable future.

By 1985 removals of coniferous sawlogs were close to 2.5 million cubic metres; in addition, nearly 1.5 million cubic metres of other coniferous wood (mainly pulpwood, pitwood, and chipboard raw material) were harvested. Removals will increase annually rising to double the 1985 figure by the year 2000.

The home sawmilling and other wood-processing industries are hungry for more wood. Nevertheless the overall production of British wood and wood products is currently meeting only about 12 per cent of the total UK demand, and imports will continue to dominate the market.

It is against this background that markets for sawn wood are considered below.

The largest demand for all types of *sawn* timber is from the construction industry which in the UK accounts for about 70 per cent of the total consumption of sawn softwoods, and 30 per cent of sawn hardwoods; hence it is this sector that offers the greatest opportunity for import substitution of sawn wood. Opportunities to increase sales to the furniture industry, which takes about one quarter of the total sawn hardwood, undoubtedly exist, but are limited by a number of factors. These include the entrenched position of tropical hardwoods — especially those such as teak and mahoganies which have a long established niche — and the lack of continuity of supplies of consistant qualities of British hardwoods in the sizes sought by furniture makers. Certain other products especially mining timber, are now supplied entirely from home sources, but these are markets diminishing in size.

The requirements for most outlets are well defined in the published British Standards. It is the policy of the British Standards Institution to move away from the traditional product specifications in which timber acceptability was defined by species, origin and visible features, towards performance specifications which do not necessarily refer to the country in which the wood was produced.

Timber frame housing

Timber frame building is a method of construction in which the weight of the house is borne by a wooden skeleton, and the entire compressive load is conveyed to the foundations by a horizontal timber base called a sole plate. It offers substantial advantages over traditional methods of building, especially in the improved cash flow resulting from the speed of erection and completion. Not only can much of the fabrication, including the precision machining of the wooden components, be undertaken in factory conditions instead of on an open building site where the rate of progress is dependent on the weather, but also the roof is usually in position at a relatively early stage thus providing shelter for interior work. Moreover, the thermal insulation of timber frame buildings is superior to traditional houses with brickwork walls.

Under this method of construction, the requirements of the Building Regulations and the National House Building Council are met by making strength calculations for individual designs because there are no "deemed to satisfy" clauses as in traditional housing. This means that timber frame houses are highly engineered with optimum use being made of stress-graded timber; nevertheless, the volume of timber used is roughly 60 per cent greater per dwelling unit, than when traditional methods are used. Timber frame designs are, however, not approved for buildings greater than three stories high.

Since 1983 the N.H.B.C. has required that where non-durable species are used for studs, rails and lintels, they must be treated with an approved wood preservative. In practice, this usually means treatment with an oil-borne preservative by the double vacuum process (see page 125).

Most of the wood used is supplied in 38 mm × 89 mm and 38 mm × 63 mm sections which have been made from 50 mm × 100 mm and 50 mm × 75 mm respectively. All major coniferous species are suitable, but machine grading is likely to be required for spruce if it is to be used in the same sizes as imported softwood. A joint study undertaken by the Forestry Commission and the Timber Research and Development Association, with the support of the home timber associations, during the early 1980s investigated in depth the suitability of machine graded Sitka spruce for studs.

This project identified twist during drying, resulting from spiral grain in the log, as the only significant problem. Strength, as indicated by machine stress-grading to the M75 category, was fully adequate with yields around the 99 per cent level; hence, there is no major problem in supplying this market with Sitka (and European) spruce cut from well selected logs, particularly if steps are taken to restrain twist during drying.

Building timber for traditional housing

The use of wood for load-bearing purposes is governed by the 1974

Amendment to the Building Regulations.* To conform with these Regulations all load-bearing members of the carcassing timber have to be stress-graded in accordance with British Standard 4978 "Timber Grades for Structural Use". The actual strength values and permissible loads are given in British Standard 5268 "Code of Practice for Permissible Stress Design, Materials and Workmanship".

Except where special designs are accepted by the controlling authority, compliance with the Regulations is achieved by satisfying the Span Tables in Schedule 6 in the England and Wales version. No current visual grade of spruce attains this requirement, which means that both species of spruce need to be machine stress-graded to at least Strength Class 3 of BS5268. Species other than spruce can meet the requirements by visual stress grading, so Scots pine supplied to the GS grade and SS grade of BS4978, and Douglas fir or Corsican pine to the SS grade would be acceptable, but much higher yields will result if machine-grading is employed. (Visually graded larch would also comply, but it is not normally used for traditional housing, because of difficulty in nailing and a reputation for exudation of resin). There is no provision for the use of the less commonly planted home-grown conifers such as western hemlock or the silver firs.

Although BS5268 permits a maximum moisture content of 24 per cent of the oven-dry weight, lower levels are desirable in these days of central-heating, especially as most of the competing softwood from Scandinavia is dried to around 18 per cent at the time of shipment and is sometimes dried to lower levels of moisture locally.

Most of the section sizes used in building are defined in British Standard 4471 "Sizes of Sawn and Planed Timber" which covers thicknesses from 16 to 300 mm and widths from 75 to 300 mm, together with a range of preferred lengths. It is self-evident that some sections such as 38 mm × 75 mm commonly specified for trussed rafters, offer a large continuing market.

Building timber: joinery

Joinery is a relatively small part of the overall requirement for building timber and it is likely to continue to be dominated by northern European softwoods and tropical hardwoods. However, British softwoods are quite suitable if selected to meet the requirements of British Standard 1186 "The Quality of Timber and Workmanship in Joinery". This Standard

* Different Regulations are in force in England and Wales, Scotland, Northern Ireland and London, but their requirements for load-bearing timber are virtually the same.

restricts the number of annual growth rings per 25 mm of radius,* prohibits dead knots and knot holes on exposed faces, and limits the size of live knots (the permissible diameter of live knots depends on the dimensions of the piece). The objective of these requirements is to ensure that timber gives a good finish, and that it works easily when cut and planed with both machine and hand tools. Other desirable features of joinery wood are stability in changing conditions of atmospheric humidity, absence of resin exudation, and resistance to splitting around nails.

Scots pine is often the preferred timber for joinery, partly because being the same species as the favoured redwood** from the Baltic region and the Soviet Union, it has comparable properties; and partly because a high proportion of the yield displays the required minimum growth rate. The relative merits of major home-grown species are given in the following table on page 98.

Improved timber structures

The use of timber as a major structural material is sometimes constrained by its inability to form a joint which will develop fully the strength of the wood, and also because of the limitations of the sizes and shapes that are commonly available.

The first of these objections, which results from the relative weakness of wood in shear parallel to the grain, has been met by the introduction of timber connectors around bolted joints, and also by the development of punched metal nail plates, now widely used in the fabrication of trussed rafters, and hand-nail plates. The second objection on limited size availability has been overcome by the development of glued-laminated beams with finger and scarf joints, which enables beams, arches and other structural units to be manufactured to virtually any desired lengths, widths and depths — often with considerable aesthetic effect (Fig 7.2).

Glued-laminated structures are built up from boards by gluing one on top of the other. Finger joints are made by the ends of pieces of wood which have been machined into interlocking finger shapes (Fig 7.3).

Laminated structures have the additional advantage that strength reducing defects such as pockets of reaction wood (see page 73) are distributed throughout the member instead of being concentrated at a single "weak point". They also have high fire-resistance — having a low co-efficient of expansion and being a poor conductor of heat, they do not weaken as quickly as steel.

* Six annual rings per 25 mm of radius for external joinery, and four for internal joinery. The figure is obtained by averaging the number of rings over a 75 mm line along the radius (where pith is present the count should start at a point 25 mm from it)

** This timber (*Pinus sylvestris*) should not be confused with that of Sequoia (*Sequoia sempervirens*) — a north-western American species which is sometimes incorrectly termed redwood.

Performance of the major softwoods in joinery

	Scots pine	Corsican pine	Lodgepole pine	European spruce	Sitka spruce	Douglas fir
Ease and quality of finishing	Good	Good	Excellent	Sometimes woolly; use of jointed cutters desirable	Sometimes woolly; use of jointed cutters desirable	Good
Likely yield from well managed plantations	High but may be reduced by large and dead knots	Low; limited by ring width, large and dead knots	May be limited by ring width*	Moderately good; may be limited by ring width	Usually low; limited by ring width	Limited by both ring width and large knots
Knots exceeding 25mm diameter	Frequent	Frequent	Moderately frequent	Rare except in edge trees	Rare except in edge trees	Frequent
Blue (sap) stain hazard	Susceptible	Susceptible	Susceptible	Absent	Absent	Susceptible
Stability in changing humidities	Moderate	Good	Good	Good	Good	Good

* Although lodgepole works and finishes well at any rate of growth (see page 145).

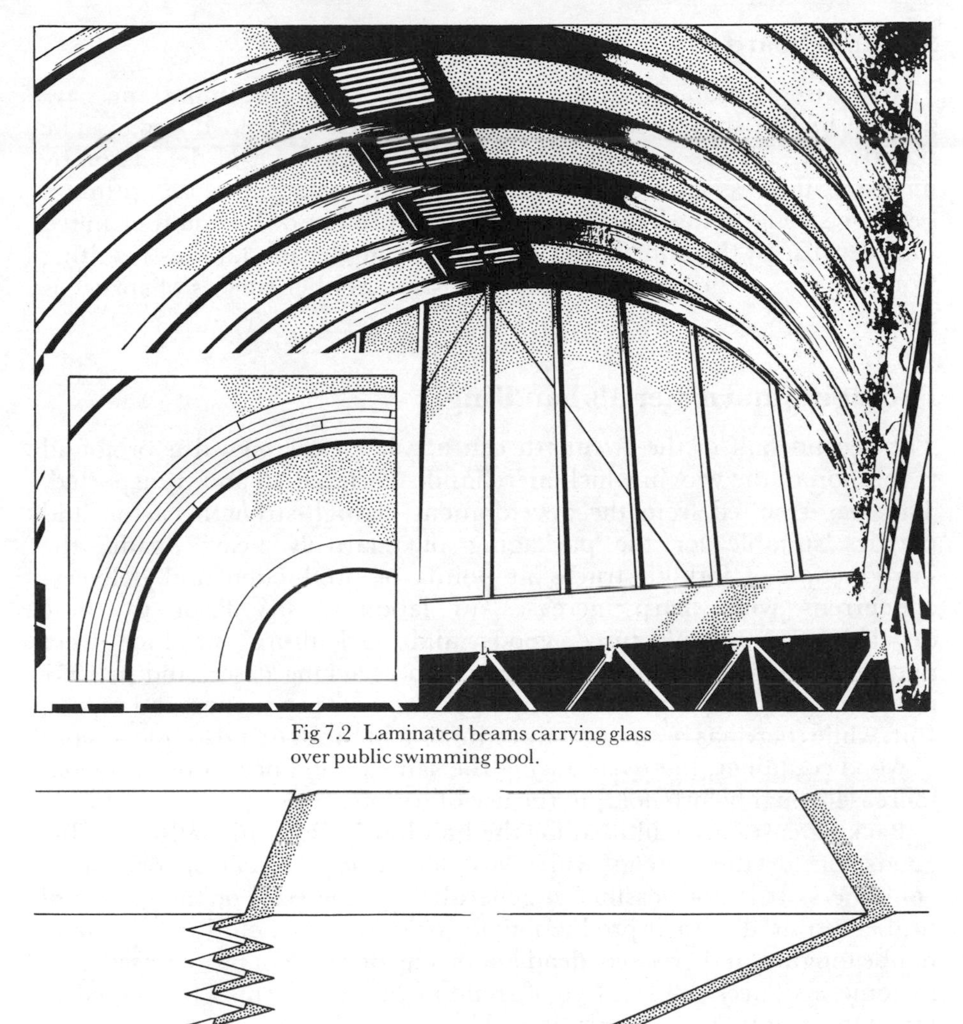

Fig 7.2 Laminated beams carrying glass over public swimming pool.

Fig 7.3 Finger joint (left) and scarf joint.

The production of laminated wood is covered by British Standard 4169,* and finger joints by BS5291,** and their application in construction work by BS5268 Pt 2.***

Laminated wood is relatively expensive, and the extent to which it is used depends largely on the price and availability of alternative structural materials.

* British Standard 4169:1970 "Glued laminated timber structural members".
** British Standard 5291:1984 "Finger joints in structural softwoods".
*** British Standard 5268: Part 2: 1984 "Structural Use of Timber, Part 2, Code of practice for permissible stress design, materials and workmanship.

Scaffold boards

The requirements for this use, where safety is of supreme importance, are defined in British Standard 2458:1981 "Timber Scaffold Boards". The boards may be selected either by a visual assessment of defects or by mechanical stress-grading. Because it virtually guarantees an optimum bending strength, the latter method is to be preferred. It should be noted that because of the difficulty in detecting compression failures resulting from mis-use of the boards on building sites, neither species of spruce is permitted.

Packaging and materials handling

The second half of the twentieth century has experienced a profound revolution in the way in which merchandise is handled and transported. This has resulted from the development of high-strength carton and plastics suitable for the packaging of relatively heavy loads, the introduction of fork-lift trucks at points of production and delivery, concurrent with sharp increases in labour costs. Prior to these developments manufactured goods and agricultural produce were normally transported in wooden boxes and packing cases, and in 1951 more than 20,000 people were employed on their production and repair. But, while there has been a drastic reduction in the demand for most types of wood container, there has during the same period been a phenomenal increase (roughly forty fold) in the use of pallets.

Packing cases are still used for the handling of exceptionally valuable goods, or for those items which are too heavy for carton or plastic containers. It is not possible to generalise on the sizes or the grades of timber required or their production; for example, the boards may or may not be tongued and grooved, dead knots may or may not be tolerated, and in some instances the surfaces of some of the boards may be planed to facilitate printing or stencilling. However, dryness to prevent the occurrence of condensation and/or mould growth is essential, and a maximum moisture content of 20 per cent is frequently specified; moreover, undried timber is of course somewhat heavier, and its use can result in higher freight charges.

Cartons and plastic boxes have almost completely ousted wood containers in the handling of fruit and bottled products. Wood is still used in the manufacture of vegetable and lettuce crates, where poplar is the preferred species because it has the advantages of light weight, a clean appearance, and ease of fastening (usually with mechanically driven staples), as well as the ease with which it can be peeled into relatively thick veneers without the need for steaming. In the production process selected logs are peeled in the cold condition into a continuous veneer between 2.5

mm and 6 mm thick, which is then guillotined into slats or boards of the required widths. This method has the significant advantage that there is no waste in the form of sawdust, and that the residual core can be sawn into quarters for the corner pieces of some types of crate. Air-drying is usually undertaken after assembly.

Pallets are a major use of British timber; both hardwoods and softwoods have been employed for all of their components, but as a general rule, softwoods are preferred for deckboards, stringers and baseboards, while specific hardwoods are preferred for the blocks (Fig 7.4).

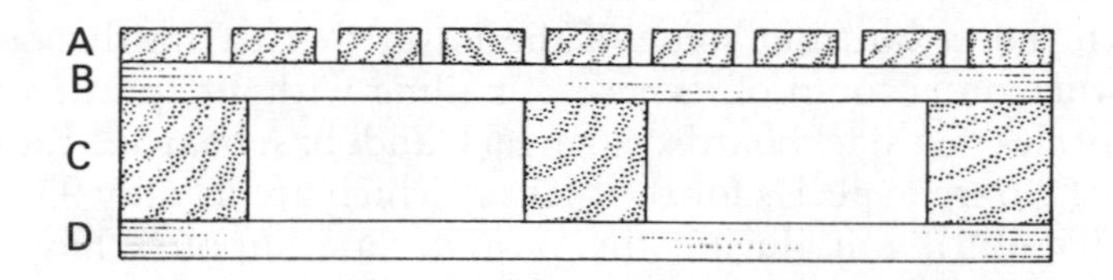

Fig 7.4 Wooden pallets. *A*, deck boards, *B*, stringers, *C*, blocks, *D*, baseboards, *E*, stack of softwood pallets.

The actual requirements for both performance and dimensions are governed by British Standard 2629 "Pallets for Materials Handling". Pallets used in the United Kingdom have traditionally been 1,000 mm × 1,200 mm, although standardisation throughout the European Economic Community might ultimately result in a preference for an 800 mm × 1,200 mm size.

While many sawmillers supply completed pallets, a much greater volume is sold as components to specialised pallet manufacturers. The greatest volume is as deckboards, which are normally 1,000 mm × 125 mm × 20 mm or 1,000 mm × 95 mm × 20 mm, and are supplied in bundles of 400 or 500 respectively. Stringers are supplied in identical sections, but are 1,200 mm in length, while baseboards are cut to both 1,000 mm and 1,200 mm lengths. Although all species of softwood other than western red cedar, and some of the lower density hardwoods are suitable, spruce on account of its ease of nailing without splitting is the preferred species for deckboards, stringers and baseboards. Elm was formerly the favoured species for the blocks (which are usually 95 mm × 95 mm × 95 mm), but during the past decade supplies have been drastically reduced following the Dutch elm disease epidemic, and in consequence blocks of softwood, ash, beech and birch are commonly used. The reason for the preference for elm was due to its cross grain which offered good resistance to nail withdrawal, shear and cleavage.

Pallet manufacture is one of the few uses where incompletely dried wood is preferred. The reasons for this concern the nailing. The undried wood is softer and requires less force to drive the nails; secondly it is less likely to split around the nail-holes; and thirdly the slight corrosion of the nails by the sap substantially increases the resistance of the nails to withdrawal. On the other hand the surface of the deckboards should be quite dry when the pallet goes into service to preclude the development of surface moulds or the dampening of the merchandise.

Railway sleepers

The technical superiority of concrete sleepers has caused the virtual elimination of wooden sleepers on main-line tracks; nevertheless there is a small but continuing demand for 127 mm × 254 mm sections in lengths of 2,590 mm for sleepers for sidings and branch lines; crossing timbers are also required in similar dimensions. Douglas fir is the preferred species, but the larches, Scots pine and Corsican pine are also acceptable. Spruce is avoided because of its poor absorption of preservatives. In heavy duty situations such as catch points and cross-overs, and in locations where the decay hazard is great, softwoods may be replaced by jarrah (*Eucalyptus marginata*) — a highly durable Australian hardwood.

Furniture

High grade hardwoods, especially oak, beech, elm and sycamore (together with some of the less commonly available species such as cherry) are in continuing demand for furniture production. The potential market exceeds the available supply by a considerable margin, and the gap is likely to increase during the next century, with the result that most furniture will continue to be made from imported hardwoods.

Pine furniture, in which smaller sound knots are used as an ornamental feature, became fashionable during the 1970's and this vogue seems likely to continue.

Although the detailed specification for pine furniture varies with the individual manufacturer, the main requirements are self-evident, namely freedom from blue stain, large and dead knots (in some situations it might be feasible to plug knot-holes), and surface pith streaks; and kiln-dried to a moisture content of no higher than twelve per cent. Selected boards of all three species of pine are suitable, but lodgepole pine will almost certainly give the best finish.

Yew is sometimes sought for veneers and for such items as cupboard door handles where they afford a pleasing contrast with darker hardwoods.

Sawn mining timber

The term sawn mining timber (Fig 2.4) includes "splits" which are softwood pit props sawn in two along their length; "crown trees" which are made from softwood roundwood cut into two along the length and then given a flat face to the resultant half-round sections by further sawing along the length, and, if necessary trimming the edges with a saw. (Sawmillers use the expression "split, backed and edged" for this product).

Square sawn timber, especially from lower grade hardwoods, is employed in a number of ways in the winning of deep-mined coal. The principal uses are for sleepers, chocks (pillarwood) for the building of temporary roof supports (Fig 2.4), lids for use on top of and under pit props, and cover boards (horizontal boards running between the arched steel girders supporting the roofs and side walls of the main galleries).

All species of hardwood other than poplar, and most species of softwood are acceptable. But, in implementing its purchases, the industry requires all suppliers to comply with the administrative procedure laid down in British Standard 5750 "Quality Systems". Intending suppliers should therefore first seek guidance from the Headquarters of British Coal.

Rural crafts

In addition to the roundwood purchased by rural craftsmen and industrial wood turners, several hundred rural-based firms buy relatively small quantities of sawn British-grown wood if it can be obtained to their exacting specifications. These include:

Type of firm	Examples of British woods used
Bar fitters	Decorative hardwoods
Beehive makers	Lime for honey frames
Boat builders	Larch for planking, oak for frames
Butchers block makers	Sycamore
Car body restorers	Ash for estate car frames
Caravan builders	Ash for framework
Carvers	Lime, oak, walnut, etc.
Furniture makers and restorers	Ash, beech, cherry, oak, pine and sweet chestnut
Ladder makers	Oak and ash for rungs
Sports goods makers	Ash (yew for archery bows)
Toy and model makers	Beech, birch, sycamore
'Tudor' framing for houses (kit makers)	Oak
Trug basket makers	Oak, hazel, poplar and willow
Turners of ornamental woodware	Most hardwoods and yew
Walking stick makers	Ash, sweet chestnut
Wheelwrights	Ash, elm and oak

Other specialist manufacturers include makers of bird mobiles; kennels; kindling wood bundles; knitting needles; maritime figureheads; musical instruments; pigeon lofts; riding boot trees; rocking horses; spinning wheels; Wendy houses; wheelbarrows including Victorian wheelbarrows; wooden jewellery; and wooden windmills.

Dunnage

Dunnage is defined in the Shorter Oxford Dictionary as "Mats, brushwood, gratings etc., stowed under cargo to prevent moisture or chafing". It does not merit a mention in the Pocket Edition.

Although the cargo which is being protected by dunnage may be on rail or road vehicle, most dunnage is (and has been in the past) used for the prevention of damage to ships' cargo. Dunnage used for this purpose is invariably some form of sawn timber. The use of dunnage is disappearing as freight is more and more frequently transported in containers, following a period when the use of dunnage was already declining as a result of the widespread use of pallets to which the merchandise was securely strapped. Prior to this era, every ship carried with it a supply of dunnage in the expectation that at least some of it would be suitable for the next cargo. Ultimately it might have to be jettisoned, either because the dunnage had deteriorated by wear or frequent wetting, or it was of a type unacceptable for a particular cargo.

Sawn timber dunnage served in a number of ways. It might, for example, be used to form a dry floor by raising the cargo on two or three layers of boarding above the ship's steel deck within a hold; or it might be used to separate two different types of freight in the same hold; or for the general protection against damage by preventing or reducing the movement of the goods in transit.

The various specification for dunnage were usually easy to meet as frequent knots, wane, tapering boards and inaccurate sawing were tolerated. Fitness for a specific purpose was ensured prior to loading by surveyors, of which there might be up to three acting for the shipper, the consignee and the marine insurance company respectively.

Although demand now is limited, dunnage still provides some sawmillers with a useful outlet for the poorer qualities of timber.

Other markets

Significant markets for sawn timber which are likely to continue and possibly increase are for portable buildings (such as temporary offices, garden sheds, and summer houses), and "Do-it Yourself" retailers. Neither of these uses have a centrally recognised specification, but for both outlets adequate drying (preferably kilning to a moisture content of 15 to 16 per cent), presentation and packaging are extremely important. Sawn wood of various grades is also in demand for many civil engineering projects, but it is not possible to generalise on the requirements.

8

Pests and Preservation

All the major components of wood tissue are subject to attack by a variety of living organisms. These range from primitive bacteria capable of slowly eroding the cell walls to elaborate higher plants such as mistletoe which invades the wood tissue of branches in search of water and dissolved mineral nutrients. However, in economic terms it is fungi and insects which constitute the most serious pests in the production and use of timber.

Fungi

Fungi can damage timber in two ways: they can cause unsightly stains and moulds, which makes it less acceptable for those uses in which appearance is important, or they can reduce the strength significantly by digesting the wood tissue, ultimately rendering it useless for any purpose other than fuel.

Growing trees are susceptible to attack by many species of wood destroying fungi some of which can persist after felling and conversion, and can continue to cause decay until the wood becomes too dry for them to flourish. In the growing tree the heartwood has a lower moisture but higher air content than the sapwood and is thus much more susceptible to attack by such fungi than the saturated sapwood. Decayed heartwood enveloped by a ring of sound sapwood is frequently manifest in species like birch or western hemlock, where it is known as heart-rot or pipe-rot, but it is virtually unknown in species such as Corsican pine or horse chestnut which are almost exclusively sapwood.

Fungal attack in the standing tree (eg by *Stereum*) can also follow harvesting damage (Fig 8.1) if sufficient bark is scuffed away from the stem to permit localised drying of the sapwood, or it can result from the

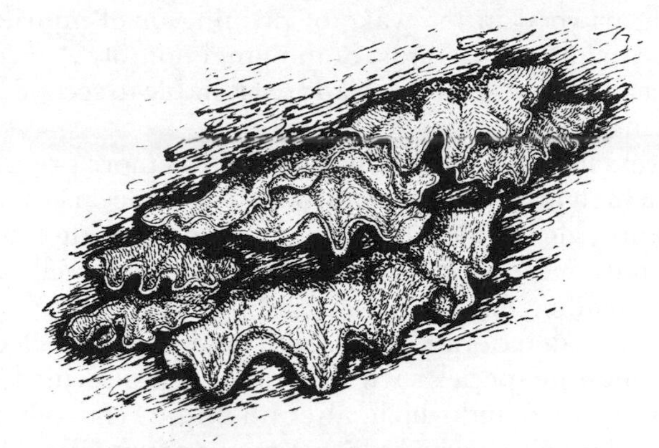

Fig 8.1 *Stereum hirsulum*. Harvesting damage to standing trees can facilitate entry of fungi via the sapwood.

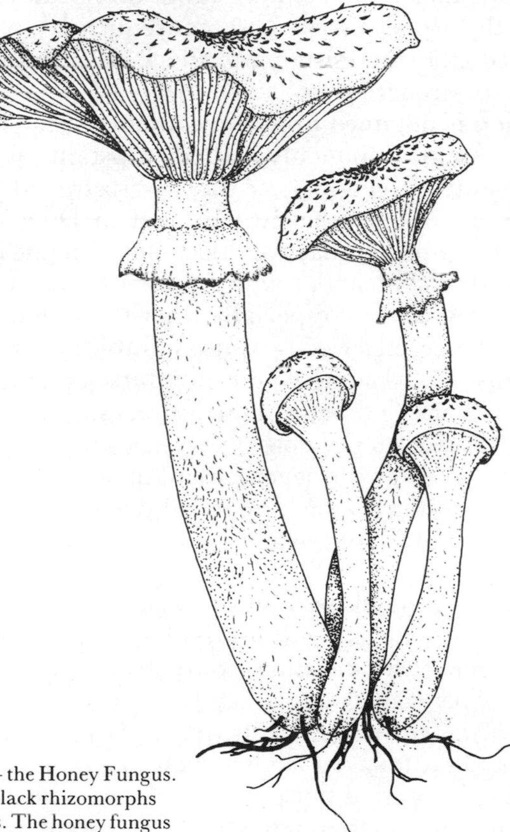

Fig 8.2 *Armillaria spp* — the Honey Fungus. Various ages showing black rhizomorphs that can penetrate roots. The honey fungus formerly referred to as *Armillaria mellea* is now known to be one of a group of similar species, with different parasitic effects.

exposure of heartwood in the wake of pruning or of injuries caused by animals such as squirrels or birds. Some fungi notably *Armillaria spp* (Fig 8.2) can kill healthy roots and the dead tree is liable to secondary attack by wood destroying fungi.

Once the trees have been felled decay becomes more prevalent in those timbers with a high proportion of sapwood. This is because partially dried sapwood can provide the ideal balance of air and moisture, together with extra nutrition from the contents of the living cells, to facilitate optimum growth of the invading fungi.

Post-harvesting deterioration by both staining and decay fungi is therefore common in species such as Corsican pine which have a high proportion of sapwood and which, after an initial relatively rapid loss of moisture, dry relatively slowly. In fact, a major advantage of felling in late winter is that the wood dries most rapidly during the following (spring) months before harmful insects and fungi become fully active.

Surface moulds and wood staining fungi do not destroy a significant proportion of the wood tissue, and consequently, with the possible exception of a reduction in resistance to shock loading, they do not cause a measurable loss of strength.

Their nutrition is obtained from the contents of the living cells of the sapwood. Blue stain, commonly called sapstain, is the type most frequently encountered (Fig 8.3). Attack is restricted to the sapwood. While it can occur in some hardwoods and in Douglas fir, it is most troublesome in the pines. It may well become epidemic in partially dried timber at a moisture content of roughly 25 per cent, during periods with still, warm, humid weather — especially when the air temperature is above 20°C. Although the affected wood is quite suitable for structural purposes, pallets and items of packaging where presentation is unimportant, it is undesirable for joinery, flooring and furniture, especially if a clear finish is to be applied. Furthermore, some purchasers take the view that if conditions have permitted the development of blue stain, such conditions could also have facilitated attack by wood destroying fungi. This is not necessarily so; staining fungi usually grow at a faster rate than wood destroying fungi.

The prevention of attack by fungi causing stains and mould is best affected by drying as soon as possible after felling and conversion with the objective of reducing the moisture content of the sapwood below the critical level of 22 per cent.

Where circumstances permit, kiln drying is to be preferred because temperatures can be attained quickly which ensure the death of the fungi. Where accelerated drying is not possible, the wood can be dipped in or sprayed with an appropriate fungicide such as sodium octoborate, or it can be immersed in or sprayed with water with the aim of raising the moisture content to levels where attack by such fungi is impossible (at

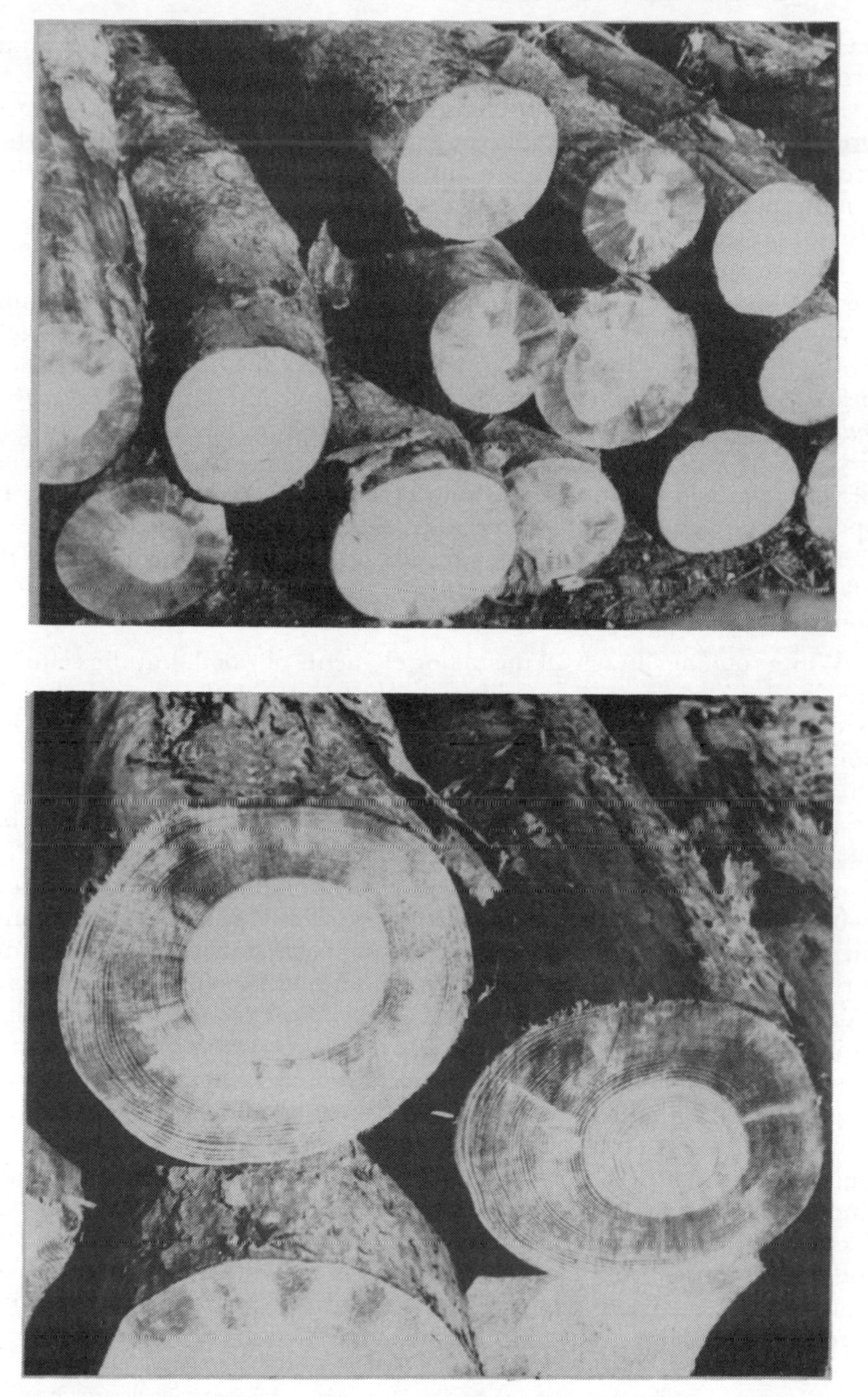

Fig. 8.3 Blue stain on Scots pine sawlogs

levels above 30 per cent there is insufficient air available to facilitate growth by staining or rotting fungi).

Attack by wood destroying fungi falls into three categories, namely brown rots, white rots and soft rots. As with stain and mould fungi they require an adequate supply of air and moisture but they differ in that they obtain their nutrition by digesting the structural components of the wood tissue and reducing the strength properties of the timber significantly.

Brown rot fungi secrete enzymes which attack the cellulose and hemicellulose of the cell walls, darkening the wood to shades of medium brown and causing the development of longitudinal and transverse cracks in the tissue. They include among many others the common household pests of Dry Rot (*Serpula lachrymans*) and the Cellar Fungus (*Coniophera puteana*) (Fig 8.4). The former is able to spread to dry wood because its mycelia growing on relatively wet wood can under certain conditions (not of very common occurrence) secrete moisture on to a new substrate. Its spores however, do not germinate on dry wood. Its comparatively frequent occurrence in buildings results largely from its tolerance of the alkalinity of mortar, which enables it to spread through brickwork, sometimes passing from one dwelling to another.

White rot fungi attack all the major elements of wood, but the cellulose may be digested much less readily than the lignin,* with the result that the decay is manifest as white axial strands. Controlled use of white rots has sometimes been suggested as a possible method of pulping wood — a process which would have a low energy requirement — or as a means of increasing the digestibility of wood waste for use as animal fodder, but neither use has been developed commercially.

Although it has been known for many decades that microfungi can grow within the cell walls of wood tissue, the destructive potential of such fungi in causing the type of decay known as soft rot was not fully appreciated until the 1940s, when they were discovered to be the major cause of early failure in the slats of water cooling towers. Soft rot is usually manifest initially as a softening of the surface layers of the wood, but it ultimately results in brash failures, the loss in strength being attributable mainly to the digestion of the cellulose chains in the secondary layer of the cell walls (see page 180). It is most likely to be troublesome in situations which are too wet to permit the development of the brown and white rots. These include fence posts in water-logged soils such as peat bogs or areas with extremely high rainfall, in boats and in timbers serving beneath the low tide mark in river and marine installations (marine fungi are of the soft rotting type). Wood which has received an inadequate loading of preservative is also prone to soft rot attack.

* There are two types of white rot fungi; one attacks all the wood components simultaneously and the other attacks the lignin preferentially.

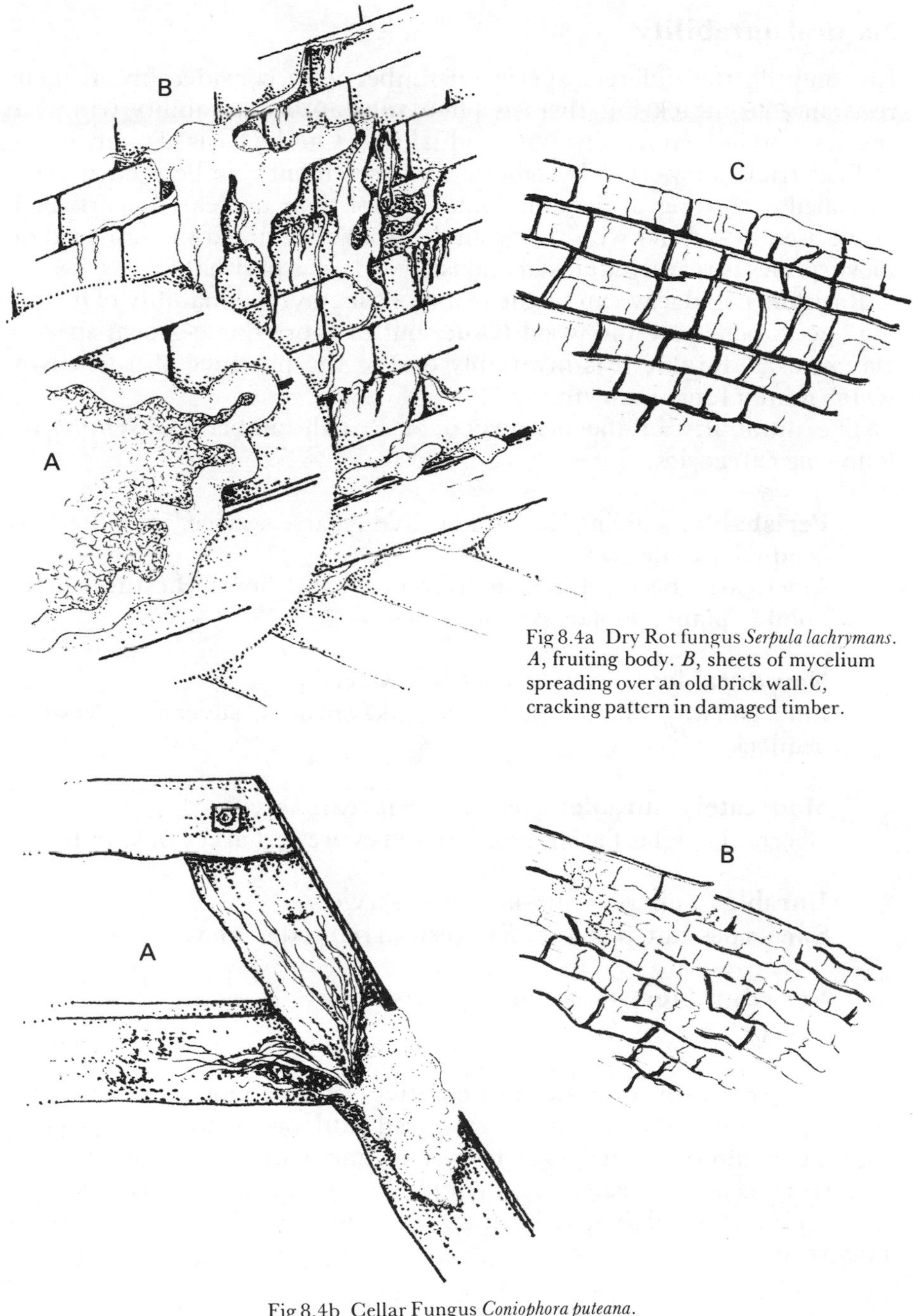

Fig 8.4a Dry Rot fungus *Serpula lachrymans*. *A*, fruiting body. *B*, sheets of mycelium spreading over an old brick wall. *C*, cracking pattern in damaged timber.

Fig 8.4b Cellar Fungus *Coniophora puteana*. *A*, mycelium and fruiting body confined to timber. *B*, cracking pattern in damaged timber.

Natural durability

Not only do the different species of timber vary considerably in their resistance to attack but there is often substantial variability within a species, and indeed within an individual tree. On the basis of many years of field trials supported by laboratory assessments timbers have been classified according to the anticipated life of their untreated heartwood when used in contact with the ground. As already noted the sapwood of most species is readily attacked and is regarded as perishable.

Resistance to decay can be the result of the physical inability of fungal hyphae to penetrate the wood tissue, but in those home-grown species classified as durable it is invariably due to the presence of extractives which inhibit fungal growth.

The durability of the heartwood of British timbers falls into the following categories:

Perishable: giving less than five years service in situations conducive to decay.
Alder, ash, beech, birch, holly, hornbeam, horse chestnut, lime, London plane, poplar, sycamore, willow.

Non-durable: five to ten years service.
Elm, Norway maple, pines, red oak, spruces, silver firs, western hemlock.

Moderately durable: ten to fifteen years service.
Cherry, Douglas fir, larches, Lawson cypress, Turkey oak, walnut.

Durable: up to twenty-five years service.
Sweet chestnut, oak, robinia, western red cedar, yew.

Very durable: exceeding 25 years service.
Leyland cypress.

Treatment with a wood preservative is recommended wherever perishable or non-durable timbers are used outdoors. Moderately durable timbers should be treated when used in contact with the ground; durable timbers need not be treated, but their performance in situations where the decay hazard is high can be improved substantially if treatment is undertaken.

Bacteria

A number of species of bacteria can digest the structural elements of timber but the rate of attack is so slow that their effects are not readily observed unless conditions of aeration and/or elevated temperature prevent the development of wood destroying fungi; they are not therefore regarded as pests. On the other hand they can be usefully employed to improve the permeability of round timber by removing the membrane of the bordered pits (see page 34).

Bacteria have also been suggested for the controlled environment delignification of wood waste to increase its digestibility for use as cattle fodder, because they are more amenable to genetic manipulation than fungi; but such a technique has yet to be developed commercially.

Insects

Insects harmful to wood (as opposed to insects harmful to trees) fall into two categories, namely those which attack the wood of living trees and fresh-felled roundwood, and those which damage wood subsequent to conversion; the former are of little economic importance and remedial measures are not required under British conditions. The latter can be controlled by sensible forest hygiene measures, e.g. removing felled wood promptly.

Freshly felled logs, and dying or debilitated trees are sometimes attacked by wood wasps — either *Uroceras gigas* (Fig 8.5) which is black and yellow and resembles superficially a hornet, or the smaller blue coloured *Syrex cyanus*. The scale of the damage which results from their wood-boring larvae, is never large. Exceptionally, the bore holes, which are four to seven millimetres in diameter and if undisturbed are tightly packed with dust (frass), may penetrate into the heartwood; and while they may look unsightly they do not cause an appreciable reduction in strength. Wood wasps do not attack dry wood.

Forest longhorn beetles (Fig 8.5), some of which have a formidable appearance, lay their eggs in round timber which has already been subject to fungal decay; however, the British species never attack sound wood.

Pin hole borers

Pin hole borers, also known as ambrosia beetles, are a pest of freshly felled wood, and they are occasionally found in standing trees. Eggs which are laid under the bark develop into larvae which bore holes up to two millimetres in diameter.

The surfaces of the holes then become covered with prolific fungal growth which is the primary food of the larvae. Although there is no appreciable loss of mechanical strength following an attack, the appearance of the wood is spoilt by stains resulting from the fungi, often extending for several millimetres around the borings. This can result in a

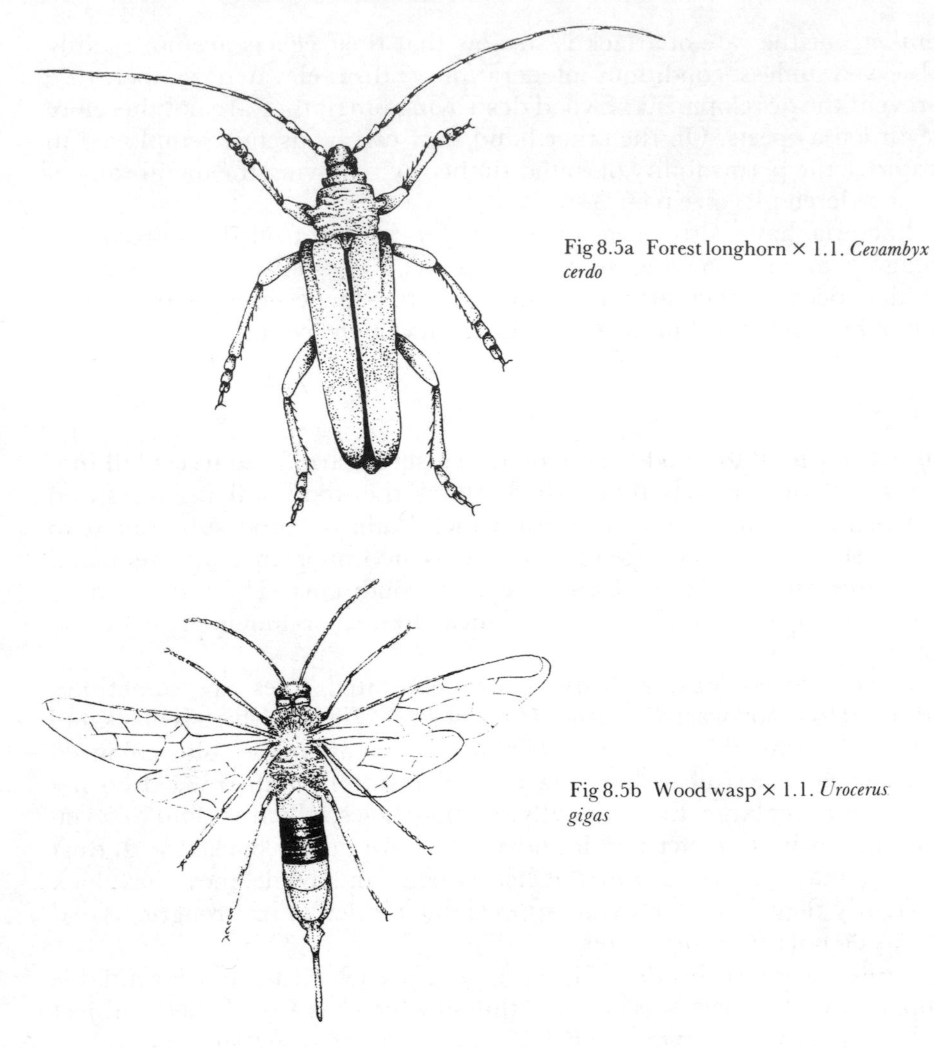

Fig 8.5a Forest longhorn × 1.1. *Cevambyx cerdo*

Fig 8.5b Wood wasp × 1.1. *Urocerus gigas*

significant loss in value in ornamental hardwoods such as oak destined for use as furniture, flooring or veneers. When the wood becomes too dry to support fungal growth the attack ceases. Preventative rather than remedial measures are required, such as restricting the harvesting of valuable logs to the winter months when the beetle is not active.

Furniture beetle (*Anobium punctatum*)

Probably the most serious insect pest of wood in use in Britain (Fig 8.6), not only because of its widespread occurence in furniture, old plywood panelling and structural timbers but also because, unlike most other

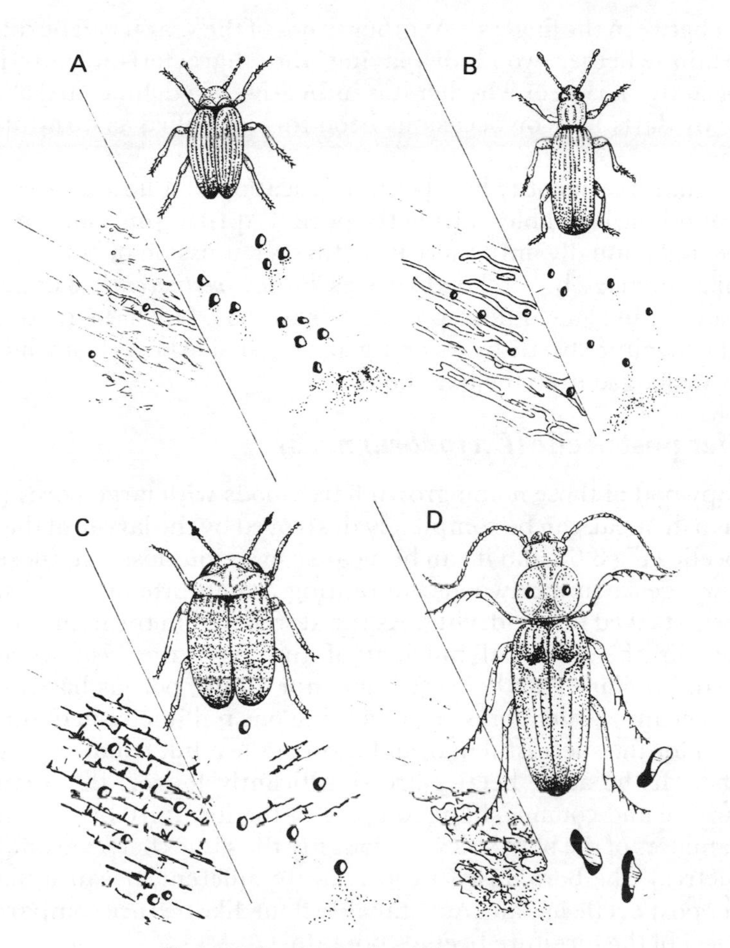

Fig 8.6 *A*, Powder Post beetle *Lyctus brunneus* × 3; *B*, Furniture beetle *Anobium punctatum* × 3; *C*, Death watch beetle *Xestobium rufovillosum* × 3; *D*, House longhorn *Hylotrupes bajulus* × 1.5

wood-boring insects which attack in the wake of fungal decay, it is able to invade both the sapwood and heartwood of absolutely sound, dry, wood. The damage is done by the minute (4–5 mm) comma-shaped larvae, commonly called woodworm, which penetrate deeply by gnawing bore-holes up to two millimetres in diameter. Ultimately, during the spring and early summer months, they return to the proximity of the surface for pupation and emergence as adult beetles, and it is then that remedial treatment is most effective. Inspection is best undertaken at this time. If the wood is tapped sharply with a mallet, bore-dust (frass) showers freely from the bore holes (the dust has a characteristic gritty feel when rubbed between the fingers). At other times of the year it can be difficult to be certain whether wood displaying the characteristic bore-holes is hosting active larvae or whether it is indicative of a defunct attack. (Holes caused by darts have on occasions been mis-identified as furniture beetle damage!).

Remedial treatment is by liberal application of a liquid insecticide in and around the bore holes during the period April to June, and this should be repeated annually until there is no further emission of bore dust.

Similar damage by a related species *Ptilinus pectinicornis* is occasionally encountered in some diffuse porous hardwoods. It can be distinguished from the furniture beetle by its non-gritty flour-like bore dust which is less readily dislodged from the bore holes.

Powder post beetle (*Lyctus brunneus*)

The sapwood of those home-grown hardwoods with large pores (vessels) such as ash or oak can be completely destroyed by the larvae of the powder post beetle (Fig 8.6), and it can be a cause of serious losses in those timber yards where sawn hardwoods containing a high proportion of sapwood are open-stacked for air-drying. As the starch within the living cells of the sapwood is the principal nutrient of this pest, the heartwood is not attacked. Similarly, neither softwoods nor diffuse porous hardwoods are at risk, because larger pores, approaching one millimetre in diameter, are required for the successful manipulation of the adult female's ovipositor.

Although the adult beetles are significantly longer, the larvae are a similar size and comma shape to those of the furniture beetle; and while the diameters of the bore holes are roughly the same (between one and two millimetres), the bore dust (frass) is quite different, because that of the powder post beetle has an unmistakable flour-like texture compared to the gritty feel of the furniture beetle's bore dust.

Death watch beetle (*Xestobium rufovillosum*)

Despite its sinister name and the publicity which it has attracted as a result of damage to our great cathedrals, the death watch beetle (Fig 8.6) is a less widespread and a less serious pest than either the furniture beetle or the powder post beetle.

Nevertheless, extensive damage can occur in ancient structural hardwoods requiring expensive replacements or remedial treatment. Attack usually follows in the wake of decay, particularly of the types which result from roof leakages; however, once the larvae are well established in pockets of rotten wood, they are able to spread to adjacent sound timber, ultimately causing the collapse of load-bearing structures. Most infestations encountered in Britain have been in oak, but other hardwoods and also softwoods may be attacked.

Both the adult and the larvae are bigger than those of the furniture and powder post beetles, consequently the bore holes are also larger being about three millimetres in diameter.

Remedial treatment is by removal of the sources of dampness in the affected building, followed by liberal applications of an insecticide. As the larvae can survive within the wood for up to a decade, repeated annual applications at the beginning of the emergence period (April–June) may be required to achieve elimination.

House longhorn beetle (*Hylotrupes bajulus*)

The house longhorn beetle (Fig 8.6) has been a serious pest of structural softwoods in domestic buildings in north western Surrey, and because of this it is also known as the Camberley beetle. It attacks sound dry sapwood.

The larvae, which bore relatively large (six to ten millimetre) holes, can bring about the disintegration of joists and roofing timbers, and in extreme cases cause their collapse.

Although the application of insecticides is virtually essential to control existing attacks, the emphasis is on prevention, especially in those areas where infestation has already been reported. In effect, this means that all softwood to be used for new structures or for repairs should be treated with an appropriate preservative, either under pressure or by the double vacuum process (local by-laws make this mandatory in high risk areas).

Other beetles

The larvae of certain other beetles are sometimes found boring into timber in use, but with the possible exception of some species of weevil, they are incapable of causing widespread damage, and are usually associated with fungal decay. The main ones include the wharf borer (*Narcedes melanura*) and *Ernobius mollis* which is encountered occasionally in the outer sapwood of timber from which the bark has not been removed completely. With regard to wood-boring weevils, there have been recent indications that the species *Europhryum confine*, *Europhryum rufum* and *Pentarthrum huttoni* may be more serious pests than formerly realised, especially in London, the Home Counties and East Anglia.

Marine borers

Unless given adequate preservative treatment, all home-grown timbers, when used in the sea are liable to be attacked by marine borers, which can eventually render them unserviceable. These comprise the gribble (*Limnoria lignorum*) and the shipworm (*Teredo spp.*) The gribble is a crustacean about six millimetres long which bores into the surface layers of timber which has already been attacked by marine fungi.*

When the outer layers of the wood have become riddled with the minute (one millimetre) bore holes they are eroded away by the action of the waves.

The shipworm is a highly specialised mollusc in which the shells are reduced in size to serve as twin cutting tools enabling it to bore deeply into submerged timber. The borings are roughly five millimetres in diameter and several centimetres long, and are lined with a protective shell-like substance.

Because of the practical difficulties involved in treatment against marine borers in-situ, and the risk of polluting the marine environment, the emphasis is on prevention rather than remedial measures, the most effective method being pressure treatment with a preservative such as creosote or copper chrome arsenate, using a schedule which will result in a high retention.

The requirements of preservatives

Preservation is the treatment of timber to render it toxic or unpalatable to those organisms which attack it. It has two aspects, *preventive* being the impregnation of or surface treatment of wood to preclude invasion, and *remedial* which involves the application of pesticides to, or the fumigation of, wood which has already been attacked.

For British timber prevention is the major facet.

Ideally, preservatives should be highly toxic to fungi, insects and marine borers, but should be non-poisonous to mammals, cultivated plants and other beneficial organisms. They should be permanent and in particular they should not change into less toxic forms with the passage of time, nor should they be readily leached away by rain or other forms of flowing water; they should be odourless and non-irritant; they should be non-corrosive to ungalvanised nails and screws and should not react with plastic fixtures; they should not reduce the performance of adhesives; and for some applications they should be colourless. No single preservative has all these features.

Coal tar creosote

Coal tar creosote has been used widely in Britain for over 150 years. Before the advent of natural gas in the 1960s it was manufactured from the tar arising as a by-product in the production of town gas, but today most of

* Marine fungi are of the soft rot type, quite dissimilar to those terrestrial species which cause the brown or white rots.

the tar is obtained from the manufacture of smokeless fuels and the production of coke for the steel industry. In the manufacturing process tar is heated to temperatures exceeding 200°C and creosote is collected as a distillate. It is a mixture of many organic compounds each of which are highly toxic to insects and fungi, and most of which are relatively insoluble in water (the proportions of the individual constituents vary with the distillation temperature and with the origin of the coal). Hence, for toxicity and permanence creosote is an excellent wood preservative, and this assertion is confirmed by century old service records. However, it has some disadvantages. Its odour makes it unsuitable for some indoor applications; oil-bound paints cannot be applied to creosoted surfaces; it "attacks" plastics; and certain species of wood destroying fungi such as *Lentinus lepidus* display a degree of tolerance to creosote and can develop in inadequately treated timber.

Its main uses today are in the treatment of fencing, telegraph and power transmission poles and railway sleepers. Some forms have a golden brown appearance and are deemed to enhance the appearance of garden fencing panels.

Copper chrome arsenate

Copper chrome arsenate is the principal water-borne preservative in use in the UK, although it is banned in some European countries. It is usually applied under pressure in an acid solution which ensures that it remains completely soluble. When the ingredients enter the sapwood the pH is elevated to levels at which they become insoluble, are precipitated and fixed within the cells of the wood tissue. Of the active constituents, copper is the most toxic to the bulk of wood destroying fungi and chromium the least. Arsenic provides the toxicity to insects and marine borers as well as certain brown rot fungi which possess a high degree of tolerance to copper.

Copper chrome arsenate does not leach out of wood to an appreciable extent, and consequently treated wood is quite safe for use in contact with food or drinking water. It can be painted over. The main disadvantage is that the treatment re-wets the timber and a second period of drying is required for most applications. As already noted on page 34 this preservative is well suited for the sap-displacement process using freshly-felled round timber and where this method is used only one period of drying is needed. Wood treated with copper chrome arsenate is odourless and is more pleasant to handle than creosoted wood.

Other preservatives likely to be used in Britain

Boron applied as a strong solution of disodium octoborate has good fungicidal, insecticidal and fire-retardant properties. It is colourless, odourless and non-poisonous to mammals. However, there is no method of fixing it and in consequence it is readily leached out of the wood. It is, therefore, quite unsuitable for external use or in any situation where repeated wetting is probable.

Copper and zinc naphthenates, chloronaphthenates, tributyl tin oxide, pentachlorphenol and sodium pentachlorphenate have been widely used for both preventative and remedial treatment, commonly as surface applications; but for the treatment of building timber before installation the double vacuum method is often used with the organic solvent (oil-borne) types of preservative.

Preservatives formulated specially for the remedial treatment against insect attack frequently contain additional potent insecticides such as lindane.

Methods of applying wood preservatives

For out-door situations where a minimum of 40 years service is required, or where the decay hazard is known to be extreme, there is almost no alternative to preservative treatment under pressure. A variety of methods, which are known by the names of their patentees, have been in operation for many decades. Some of these are called full cell processes as they aim to leave as many cell cavities as possible filled with preservative (usually creosote). Such methods are particularly suitable for the treatment of species like spruce or Douglas fir which are resistant to impregnation and where a radial penetration of only a few millimetres can be anticipated; or for timber to be used in environments where the hazard is exceptionally high, such as piling in water where marine borers are known to be active. Full cell processes on the other hand are not economic for permeable timbers such as Corsican pine, because too much preservative is absorbed; and there may also be trouble with "bleeding" i.e. the exudation of creosote after installation, which in situations such as telegraph poles can lead to compensation claims in respect of soiled clothing.

The alternatives are the empty cell processes, where the objective is merely to coat the cell walls with preservative, recovering that which had already occupied the cell cavities. It is most suitable for permeable timbers such as Scots pine and birch.

Incising

The penetration of preservative into resistant timbers can be improved by passing it between spiked rollers to create shallow incisions at regular intervals. This technique has been widely practised in North America, but is less commonly used in Britain.

The individual processes are:

Bethel (full cell for dry timber) (Fig 8.7) The wood is loaded into a pressure cylinder and subjected to an initial vacuum of about 0.6Kg/cm^2 to create suction within the cell cavities. The cylinder is then filled with hot creosote* and pressure at about 9Kg/cm^2 to 10Kg/cm^2 is applied for up to one hour forcing the preservative into the wood; and finally a further vacuum is applied at about 0.6Kg/cm^2 for about an hour to dry off the surface.

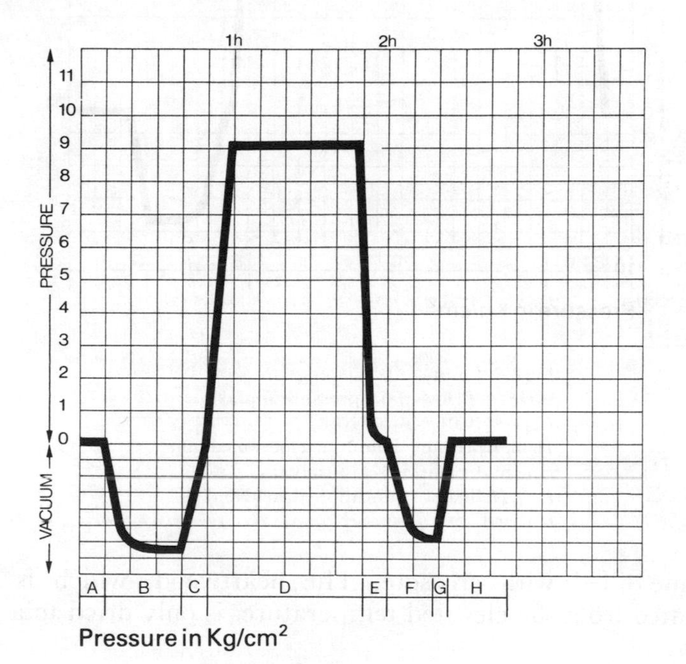

Fig 8.7 Full cell (Bethel) pressure cycle. *A–B*, initial vacuum applied; *C*, preservative introduced; pressure increased; *D*, maximum pressure held; *E*, pressure released; preservative drained out of cylinder; *F, G, H*, Final vacuum applied and released.

Boulton (full cell for green or partially dried wood) This process is also known as boiling under vacuum. Undried timber is placed in the pressure cylinder which is then filled with creosote at a temperature of around 100°C and a vacuum is applied for several hours during which time much of the water in the sapwood boils away. Subsequently pressure is applied and the cells of the sapwood

* In this and other processes creosote is normally heated to reduce its viscosity.

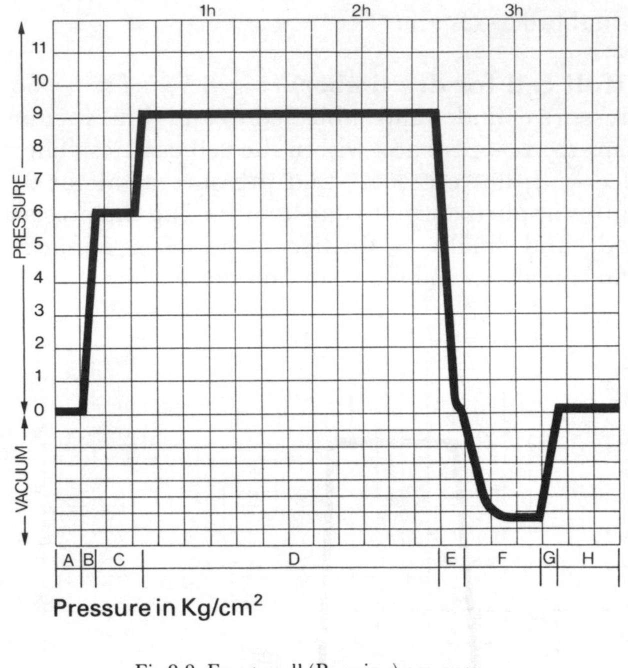

Fig 8.8 Empty cell (Rueping) pressure cycle. *A*, *B*, low initial pressure applied; *C*, preservative introduced; *D*, high pressure maintained for some hours; *E*, cylinder emptied of preservative; *F*, *G*, *H*, final vacuum applied, held and released.

become filled with creosote. The heartwood, which is largely insulated from the elevated temperature, is only dried to a limited extent.

Rueping (Empty cell for dry timber) (Fig 8.8) After the timber is loaded into the cylinder, a low intial pressure of about 6Kg/cm^2 is applied for roughly one hour to compress the air within the cell cavities, after which the cylinder is filled with preservative at a higher pressure of about 9Kg/cm^2 to 10Kg/cm^2 and this is maintained for a few hours. Finally a vacuum of 0.6Kg/cm^2 is applied. Both the initial pressure within the cells and the final vacuum help to remove and recover the excess preservative.

Lowry (Empty cell for dry timber) (Fig 8.9) After loading with the wood the cylinder is filled with preservative at atmospheric pressure, and then the pressure is raised to 7Kg/cm^2–10Kg/cm^2 for a few hours, after which a final vacuum is drawn at about 0.3Kg/cm^2.

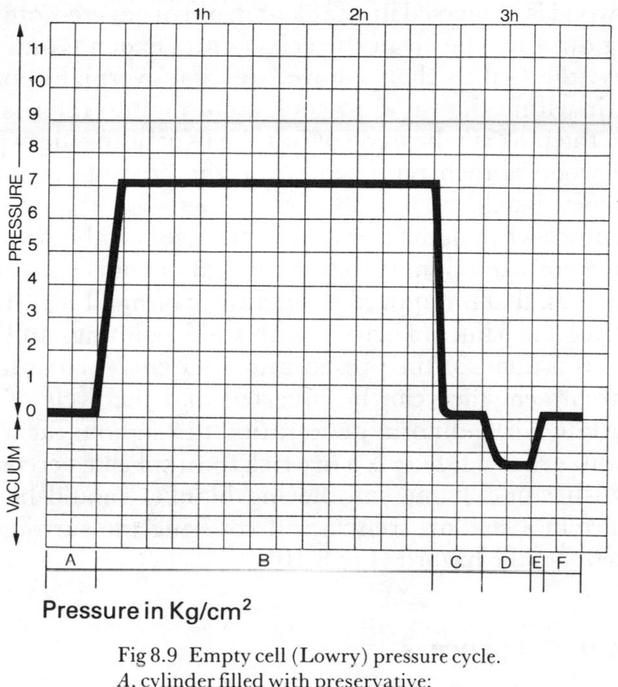

Fig 8.9 Empty cell (Lowry) pressure cycle.
A, cylinder filled with preservative;
B, pressure built up and held for some hours;
C, cylinder emptied of preservative as pressure reduced;
D, E, F, final vacuum applied for short period.

In the above processes, the temperatures, pressures, degrees of vacuum, and lengths of the various stages of treatment can be adjusted to suit the dimensions of the timber, to take account of its resistance to impregnation, or to attempt to achieve a desired specification; with low density species particular care should be exercised because high temperatures and pressures may cause collapse of the wood tissue. There is, however, a pressure treatment method which employs low temperatures, namely the Drilon process in which a preservative such as pentachlorphenol or tributyl tin oxide, normally applied by organic solvents, is dissolved in liquified butane gas. Good penetration is achieved and there is, of course, no rewetting of the timber.

Open tank (hot and cold) treatment with creosote

Adequate treatment for fence posts and other estate timbers can be undertaken by the open tank process. It is an excellent "Do-It-Yourself" method for farmers, because little special equipment is required and skilled operators are not needed.

The dried wood is immersed in a tank or drum of hot creosote for several hours, during this time the air in the cell cavities expands and much of it bubbles away; the tank is then allowed to cool overnight enabling the contracting air within the wood to suck creosote into the tissue; on the following day the tank is reheated so that the expanding air expels excess creosote, the wood is then removed while the creosote is still hot and replaced by a new batch.

While this process can be accelerated by the use of elaborate equipment such as heating coils and pumps to transfer hot or cold creosote from one tank to another, all that is required to operate on a small scale is an old oil drum mounted on two lines of bricks, with stake pointings and off-cuts as fuel. Where this is done for the preservation of items such as fence posts, after initial treatment they can be inverted and the cycle of treatment repeated to attain fairly uniform penetration throughout the length. The main disadvantage is that there is a fire risk from splashed creosote, hence a fire-extinguisher or a flame smothering blanket should be available. Putting the fire in a shallow trench and covering the surrounding area with turves reduces the hazard (Fig 8.10).

Sap displacement process

Details are given in Chapter 2 (page 34).

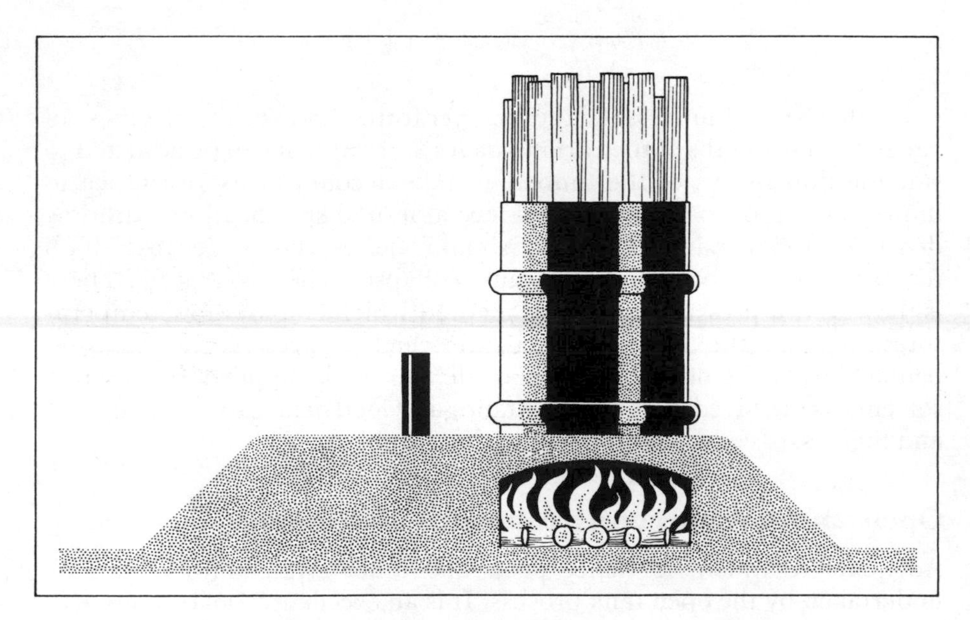

Fig 8.10 Creosoting posts in 40 gallon drum.

The double vacuum process (Vac vac)

This process was developed during the past two decades to improve the performance of external joinery timbers following unsatisfactory results given by dip treatments using organic solvent preservatives. It is now widely used by manufacturers of wood building components, especially for timber frame housing. It has been approved by the National House Building Council and certified by the British Board of Agrément.

The wood is loaded into a rectangular chamber (Fig 8.11) and a vacuum of minus 33.65Kg force per cm^2 (minus 0.33 bar*) is applied for three to ten minutes. An organic solvent preservative such as pentachlorphenol is then introduced and the wood is soaked for periods of up to one hour (with the more impermeable timbers additional pressure is exerted), after which the preservative is drained away and a final vacuum applied for about twenty minutes.

The method has the advantages of attaining much better retentions and penetrations than can be achieved by surface treatments, and a relatively short treatment cycle of between one and two hours (the actual

Fig 8.11 Double Vacuum unit.

* 1 bar = 101.97 Kg force per cm^2.

time depending on the species of timber, the dimensions and the pressure applied).

Diffusion processes

A main role of sapwood in the growing tree is the conduction of water from the roots to the crown, and to fulfil this function it has to be completely permeable to fluids. However, when some species such as Douglas fir are dried, the pits through which the water passes, in the growing tree, close completely and the wood becomes quite impermeable. The diffusion process takes advantage of the free transfer of moisture between cells in green timber. The method is to apply a strong solution of an appropriate preservative such as disodium octoborate to sawn wood by immersion or by spraying, followed by close-piling with a plastic cover to minimise evaporation for a number of weeks, after which the wood can be open-piled for drying.

Satisfactory penetration (including the heartwood) of resistant timber such as Sitka spruce can be achieved after a diffusion period of eight or nine weeks provided that the initial moisture content exceeds fifty per cent of the oven dry weight. It is interesting to note that in the green condition, spruce is more permeable than pine; complete penetration by diffusion occurs about two weeks earlier in the sizes commonly used for building.

As already noted on page 36, a two-stage diffusion treatment of fresh-felled poles is feasible using copper sulphate, arsenic pentoxide and ammonia solution, but it has not yet been practised commercially.

Surface treatments

Surface treatments including brushing, spraying and steeping are widely practiced, generally using organic solvent preservatives such as naphthanates and pentachlorphenol, or creosote. While these methods undoubtedly improve the service life of the timber, only limited penetration may be attained and hence they are somewhat less suitable for situations where the decay hazard is high. Many of the proprietary preparations include tints to give the wood an attractive colour as well as extending its life.

9

Fuelwood, Charcoal, Wood Gas and Fire-resistance

It is a profound paradox, that while wood is likely to remain a major fuel in the foreseeable future, particularly in Third World countries, it is nevertheless one of the most fire-resistant structural materials. Although roughly 98 per cent of the oven dry wood substance of all home-grown species is readily combustible, a well-designed timber structure will remain standing in a fire of several hours duration, while certain other non-combustible construction materials, under similar conditions, are likely to experience a drastic loss of strength which could lead to complete collapse. Many failures of timber structures in recent years have, in fact, originated from underlying accumulations of rubbish or the combustible contents of the building in question, especially where such buildings were erected before the principles of fire-resistance were understood. This apparent contrariety can be explained by considering the behaviour of wood at elevated temperatures.

When wood is subjected to temperatures not exceeding 150°C there is merely a release of moisture without any significant thermal degradation of the tissues. Above this level inert gasses such as carbon dioxide are emitted; but it is not until the temperature exceeds 250°C that there is any appreciable release of flammable gases resulting from the decomposition of hemicellulose, at which point ignition may be possible. This is followed by breakdown of the cellulose at temperatures around 300°C and finally of lignaceous compounds at temperatures above 320°C. Although complete combustion can be achieved at temperatures between 300 and 400°C, higher temperatures are necessary for the production of gases rather than liquids, and to ensure the volatilisation of tar. The gases liberated include hydrogen, carbon monoxide, carbon dioxide, and methane, together with smaller quantities of other hydrocarbons. It is the ignition of these gases (other than carbon dioxide), usually by adjacent combustible material, which causes flames to appear on the surface of the wood; even so, the rate of emission of such gases may not be sufficient to allow sustained burning.

Both wood and charcoal are poor conductors of heat and so little of the heat produced by the combustion of the flammable gases reaches the inner layer of wood and they remain below the temperature required for thermal degredation to take place. It is therefore clear that for optimum fire resistance the ratio between the exposed surface and the volume of the piece should be kept as low as possible. In contrast kindling wood, with individual pieces of small cross section, has a high surface to volume ratio which insures continuous burning.

Fuelwood

Wood is still a major source of energy in the developing world, and indeed, for the poorest families, wood gathered without payment is the only source of domestic energy available. The use of locally harvested fuelwood in the poorer rural areas of the developed world began to decline significantly with the advent of distribution networks for bottled gas and for oil and with the increased availability of electricity in rural areas.

In Britain, the widespread availability of coal together with the intensive rural electrification programmes of the immediate post-war years allowed wood to compete successfully only in rather exceptional situations such as in remote parts of Scotland distant from railheads and ports, but near to forests. Nevertheless the convenience of electricity and bottled gas and fuel oil resulted in a gradual erosion of the fuelwood market, so that by the late 1960s little wood was used as domestic fuel.

In industry the use of wood as fuel in the two decades up to 1965 was confined almost entirely to the combustion of residues in sawmills and woodworking plants, where the main uses were in:

- Portable sawmills belt driven by steam traction engines fired by slabwood and offcuts.
- Permanent sawmills, electrically powered from the mains, but using residues to raise steam for timber drying kilns, space heating, and to generate a supplement to the electricity supply at peak periods.
- Wood-working plants using sawdust, shavings and off-cuts for space heating, or less frequently to generate electricity.

One or two pulpmills burned the bark from the pulpwood billets which had been debarked at the mill, especially where other disposal methods were more costly.

Alternative and more remunerative uses for residues were, in fact, increasing. The manufacture of wood chipboard (particle board) continued to expand, and those sawmills with efficient debarkers were able to market bark-free softwood residues to new production units. A few sawmills converted their bark-free softwood residues to chips for sale to pulpmills at home and in Scandinavia.

As in Europe generally, the role of wood as an industrial and domestic fuel continued to decline into the 1970s in the face of competition from coal, oil, electricity and natural gas, until a dramatic increase in the price of oil in 1973 stimulated a new interest in the use of wood as a domestic fuel. The sale of Scandinavian style enclosed wood-burning stoves soared, and although wood is not important in the overall use of domestic energy, in some areas firewood now competes in price with pulpwood — especially broadleaved pulpwood. This has had a beneficial effect on the management of broadleaf high forest, stimulating the demand for the poorer qualities and the smaller sizes of thinnings. Similarly, as a result of the demand for firewood, the management of some woodland areas as coppice has remained possible, even though the traditional markets for their produce, associated with underwood craftsmen, have largely disappeared. In the more densely populated regions it is sometimes possible to sell early conifer thinnings as firewood.

Production of fuelwood based on roundwood removals from the forest

Thousands of cubic metres

	1950	**1970**	**1975**	**1980**	**1985**
UK Broadleaf	—	206	80*	80*	71
UK Coniferous	—	186	30*	30*	58
Total UK	433	392	110*	110*	129
Europe Broadleaf	—	53,033	40,385	37,082	44,095
Europe Coniferous	—	18,245	15,364	18,015	18,918
Total Europe	121,800	71,278	55,749	55,097	63,013

* estimates

Sources: FAO Yearbook of Forest Products Statistics; Forest Products Statistics ECE/TIM/31 (UNI986); Timber Bulletin for Europe Vol XLI No 4 Geneva 1988.

For industry on the other hand, the use of wood as a source of energy did not increase significantly in response to the abrupt rise in oil prices. There are several reasons for this:

- a reluctance to invest in new fully automated wood-burning furnaces together with the equipment required for storing and chipping the fuelwood (it would have been quite unrealistic to have used traditional labour-intensive methods).
- the increased demand for sawmill and other wood-working residues by the particle board industry, and to a lesser extent by the pulp mills.
- the reduction in oil prices which occurred in the early 1980s just as

wood fuel technology had improved to the point of offering a more attractive alternative to other fuels than it had in the 1970s.

There are no reliable statistics on the use of wood as fuel in the United Kingdom, but the *trend* in production of fuelwood, quantified as removals from the forest is shown in the table above, together with those for Europe (including eastern Europe).

The efficiency of wood as a fuel

Fresh felled wood contains a substantial amount of water which can vary from 45 to 200 per cent of the oven dry weight of wood, depending on species and the proportion of sapwood. For efficient burning most of the water must be removed, otherwise it will impede the access of oxygen, and an appreciable proportion of the warmth produced will be lost as latent heat; moreover the combustion of damp wood can result in an excessive deposition of tar in flues, and in an unacceptably large emission of smoke.

Adequate drying can be achieved by stacking the wood in the open air or in a well-ventilated shed in such a way as to allow air circulation around the individual pieces. Tightly packed wood will dry only very slowly — so slowly that decay may occur with consequent loss in calorific value. Air drying down to 20 per cent is recommended, and this may well take up to a year, although for smaller pieces, sufficient drying can occur between the months of March and July, which is the best period for drying under UK conditions.

The calorific value of air-dry wood is between 16,000,000 and 18,000,000 Joules per Kilogramme; coal at 30MJ/Kg has nearly twice that calorific value, and fuel oil at 40 MJ/Kg has two and one half times the calorific value of wood.

A major disadvantage of wood as a fuel is its bulkiness. This does, of course, vary with the species. Lighter woods such as spruce and poplar occupy more space than heavier species like oak or beech. An average figure quoted by the Forestry Commission is that one and three quarter tonnes of air-dry wood when stacked occupies about four cubic metres, and this will provide approximately the same heat as one tonne of coal which occupies a mere three-quarters of a cubic metre.

Prospects for the increased use of wood as a source of energy

Vital though wood is in the developing world as a source of energy, in Europe fuelwood and wood residues, together with the energy obtained by burning waste liquors in pulpmills, account for roughly one half to one per cent of the total energy consumption. Within this general perspective domestic firewood may be important locally, and in the forest industries of the ECE region (Europe, USSR and North America) wood is a significant source of energy; indeed waste liquors and solid wood residues account for about one quarter of the energy requirements of the forest industries as a whole. Nevertheless there is no serious possibility of extending the use of wood as an energy resource to any appreciable extent — for example, if the whole of the annual increment of the stem wood in the ECE region were used for energy it would provide only ten per cent of the energy needs of the region.

While there is no general prospect for a much greater use of wood for energy, it is possible to envisage situations where forest residues, early thinnings and "energy plantations" of fast-growing coppice could compete with fossil fuels to provide local industries with energy. In Britain this would mean that areas distant from the coal fields and the gas grid, as well as from ports with facilities for handling fossil fuels in bulk, would be suitable for the establishment of "energy plantations" using fast-growing crops such as *Eucalyptus spp*.

It is commonly believed that in the event of another energy crisis, wood might have an important role to play as a substitute for fossil fuels. This is only true when the wood has no alternative uses. A far greater contribution can be made to closing the "energy gap" by using wood and wood products to replace other materials which have a much higher energy requirement in their manufacture. For example, if solid wood be used instead of steel or aluminium, the net saving in energy would be greater than that produced by utilising the wood used in the substitution as fuel. (According to Haseltine the energy requirement for the production and transport of steel is 227 times greater, and aluminium 650 times greater per cubic metre than that of solid wood).

Charcoal

The use of charcoal for domestic cooking is probably as old as the use of fire. The use of charcoal for industrial purposes is at least as old as the use of metals; in Britain, evidence of its use goes back to 2,500 BC. Although coke was used for smelting iron as early as 1708, by Derby at Coalbrookdale, it was only during the 19th century that charcoal was replaced by coke as the main smelting agent. It is still used to produce certain top-grade steels because of its low phosphorus and sulphur content.

The main uses in Britain today are for the refining of non-ferrous metals, the manufacture of alloys, together with annealing and other special heat-treatments; fuel for barbecues; in agriculture, horticulture and animal feeds: and in the activated* form for the refinement of chemical solutions, for water-filtration and in electrical batteries; as an

* Activated carbon is made by subjecting charcoal or similar products to high temperatures together with low-pressure steam. This alters the structure to bring about a five-fold increase in micro-porosity. However, little activated carbon is today made from wood charcoal. Mineral coal and carbonised nut shells are now the preferred materials.

absorbent for gases in for example acetylene cylinders and light-buoys; and in the pharmaceutical industry, especially for stomach disorders and wound dressings (in the mid-19th century army and navy surgeons made poultices from charcoal mixed with bread crumbs for the treatment of ulcers and gangrenous sores).

In many parts of the developing world charcoal remains an important domestic fuel particularly in cities where its light weight relative to its heat output makes it a more practical alternative to wood in terms of transport costs and storage space. It is still the main source of energy for smelting in some developing countries.

Charcoal is made by heating wood in the almost complete absence of air so that the wood does not burn but breaks down chemically. The volatile products — tars, oils and naphtha — are driven off leaving a residue of carbon and a little ash. Once it has started the process is exothermic, i.e. it produces more heat than it absorbs; for that reason only a small initial fire is needed to convert a large quantity of wood into charcoal.

In the forest, woodland depot or timber yard portable kilns are sometimes used for charcoal production. These are typically vertical metal cylinders about two metres in diameter and approximately the same height, made up of two interlocking rings (Fig 9.1). The entry of air into the kiln is regulated by movable rectangular tubular vents, to which chimneys can be attached. They are placed beneath the lower ring, with sand or loose soil providing an air-tight seal. The kilns are filled with wood billets, cleft wood or off-cuts, leaving a small hole in the centre of the pile. A metal lid, which may or may not have closable vents, is positioned over the filled cylinder.

The charge is then fired by igniting tinder at the bottom of the kiln and subsequently covering the access hole with sand or soil. The ignition results in a column of hot gases rising up the centre, which is then drawn across the kiln and downwards by the chimneys. By this method larger pieces of charcoal can be produced than by burning the charge from the lower levels upwards.

It takes several days for the wood to be converted into charcoal during which time the operator must be in regular attendance to monitor the rate of carbonisation and to adjust it by regulating the entry of air. Experience rather than instruments guides the charcoal burner.

The volatile products are lost into the air through the chimneys and vents. These can be corrosive and they make the use of portable kilns near dwellings quite unacceptable.

Permanent kilns of steel or masonry, operating on the same principle, are occasionally found attached to sawmills, but they have almost disappeared in Britain. More efficient than the simple charcoal kiln are retorts or continous kilns where the volatile products are recovered or are fed into a furnace to pre-dry the wood before it is introduced into the retort. Retorts have the advantage of not discharging undesirable gases into the environment and they are therefore acceptable in urban situations. There are no longer any continuous kilns nor retorts operating

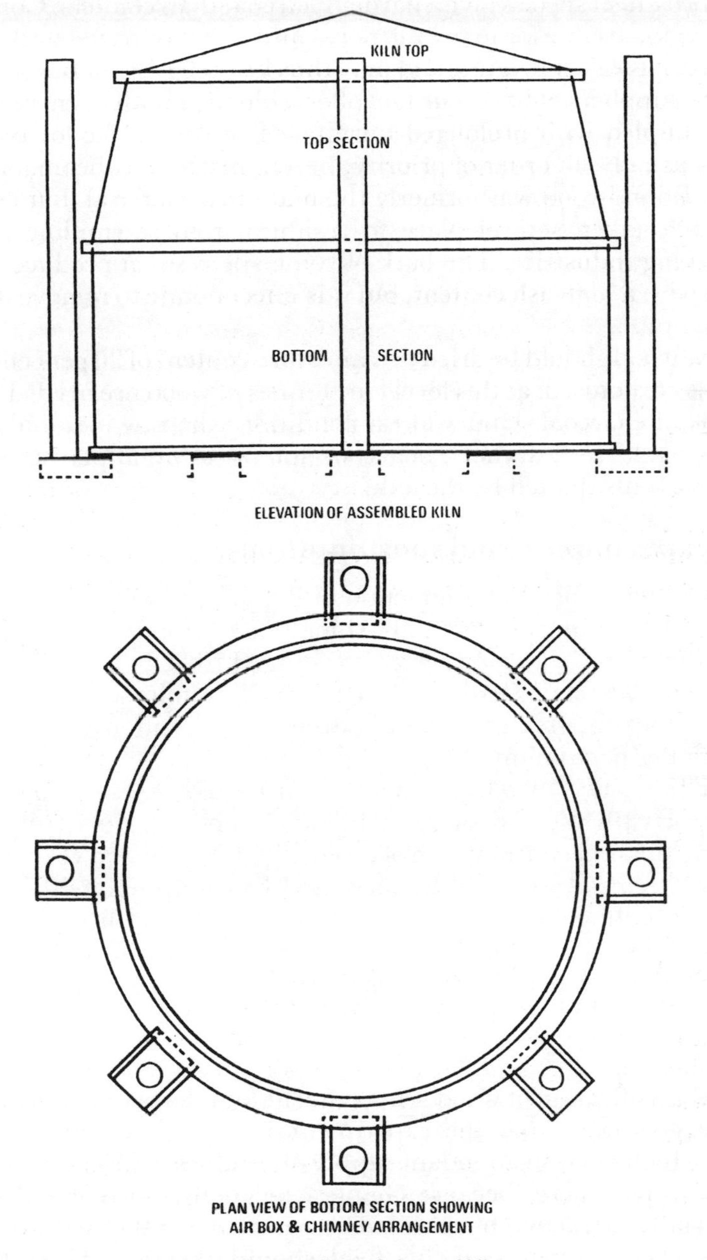

Fig 9.1 Portable steel charcoal kiln.
Courtesy Carbotech

in Britain, largely because the low cost of imports and of synthetic products derived from petroleum have made the recovery and refining of the by-products uneconomic.

Most broadleaf species give a harder charcoal than conifers. Coniferous charcoal is too friable for many purposes and may not travel well. On the other hand it is easier to ignite and may thus be preferred for barbecues. In fact, some suppliers of barbecue fuel offer a blend of both to ensure ease of ignition coupled with prolonged burning. The preference for broadleaf species is usually, in order of priority, beech, birch, hornbeam, oak, ash, and elm. Roundwood was formerly the main raw material, but in recent years it has taken second place to residues from sawmilling and the woodworking industries. The bark of some species can produce inferior charcoal with a high ash content, but it is uneconomic to remove it before carbonisation.

Ideally, wood should be dried to a moisture content of 20 per cent or less before it is carbonised; at this level four tonnes of wood are needed to yield one tonne of charcoal. Under forest conditions the raw material seldom achieves this level of dryness, hence six tonnes of wood per tonne is the figure commonly quoted by the industry.

Production, imports and consumption

In the late 1950s charcoal production in Britain was of the order of 8,000 tonnes a year, requiring annually some 53,000 tonnes of wood, ie 6.6 tonnes per tonne of charcoal. Imports averaged 16,892 tonnes a year in the same period. Deducting exports of just over 610 tonnes a year and adding home production, this gives an apparent consumption of a little over 24,000 tonnes per annum.

Production since 1980 has been of the order of 5,000 to 6,000 tonnes a year; imports 19,400 tonnes a year (1980–83); and exports virtually nil — giving an apparent consumption of around 25,000 tonnes a year. Thus the trend has been towards slightly increased consumption, with the home production falling in the face of increasing imports from developing forest-rich countries, where labour costs per tonne are lower.

By 1989 there were fewer than ten home producers of wood charcoal.

Gasification

While the gasification of wood is not currently undertaken in Britain, and has barely moved from the experimental stage in countries such as Canada which have an abundant supply of residues, a brief account of the process is given here, because changes in energy costs could make it commercially attractive in those areas where natural gas is not readily available. Developments in this field should therefore be kept under review.

It has already been noted that during the carbonisation of wood for charcoal manufacture, under British conditions, the volatile products (other than those which burn during the process) are relinquished by allowing them to escape into the air. However, in gasification, which takes place under similar conditions of restricted air availability, the charcoal is "sacrificed" to generate heat for the process, or allowed to oxidise into carbon monoxide, and the volatile products, after some degree of cleansing, are piped away for use as fuel.

In the gasification process carbonisation takes place in an air-tight chamber in which the raw material, such as chipped lop and top, can be continuously fed by a screw mechanism, and into which a controlled supply of air can be blown. There is also provision for the continuous removal of ash.

The actual composition of the product depends on the nature of the raw material, in particular its moisture content and the temperatures attained. The gas after cleansing comprises mainly carbon monoxide, carbon dioxide, nitrogen and methane in varying proportions. If the moisture content is too high or the temperature of carbonisation too low, deposition of tar can be troublesome (as already noted temperatures above 400°C are required to volatilise tar).

Gasification appears to offer the following advantages:

1. The feedstock does not need to be "pure". Forest or sawmill residues could be mixed beneficially with coal, lignite, peat or organic waste.
2. Impurities in the feedstock in small quantities are of little consequence.
3. Unlike the production of charcoal, the moisture content is not critical. Provision can be made for the drying of fresh residues within the carbonisation chamber.

On the other hand the gas produced has a relatively low calorific value. Also provision has to be made for the cleansing and for the removal of moisture from it before combustion, and this can entail an effluent disposal problem. The ash which is also produced has some value as a potassium-rich fertiliser.

Wood and wood-based panel products in fire

In global terms about half the annual harvest of wood is used as fuel, mostly in the poorer countries of the Third World. Yet wood properly used as a building material can maintain its load-bearing capacity longer than structures of non-combustible materials such as steel or aluminium. At its simplest the two extremes are often seen in the aftermath of forest fires where the twigs, branches and foliage are totally consumed, but the tree trunks and larger branches are left, charred on the outside, but still standing. This is because the thermal insulation properties of wood are high enough to ensure that the tissue only a few millimetres inside the burning zone of a tree trunk or a sawn joist is merely warmed, and becomes nowhere nearly hot enough to ignite. In contrast, structural materials of high thermal conductivity heat up uniformly and this can lead to a loss of strength throughout the section resulting in its collapse.

Fires in buildings stem from many causes, but it is the contents of the building at the time of the fire which frequently play the dominant role in determining how quickly the trouble spreads and the amount of heat generated as well as the temperatures attained. Unfortunately, new lessons are being learned from accidental fires in the home, industrial buildings and other structures irrespective of whether they contain wooden or wood-based components. The regulations governing the design of buildings, and the use of materials in their construction and furnishings are under constant revision in the light of experience, since it is not possible to simulate all the combinations of the relevant factors and their interaction in the laboratory.

The notes below on the behaviour of wood and wood-based panels in fire are an inevitable simplification of a complex phenomenon, and the references to British (and other) Standards need to be checked for changes resulting from new knowledge.

It is convenient to discuss the behaviour of a building material during the first stages of a fire (reaction to fire) and its contribution to structural stability when it is exposed to a fire that has developed fully (fire resistance).

Reaction to fire

Reaction to fire is assessed by a number of tests.

Non-combustibility

Wood and its derivatives are classified as combustible. Although the British Standard and the International Standards Organisation tests for non-combustibility give only a rather crude indication of its behaviour in fire, in some instances codes of practice are based on the classification derived from these tests.

Ignitability

In the ignitability test, wood and wood-based panels in the thicknesses used in buildings are classified as not easily ignitable. For this test a panel of the material is exposed to a small source of ignition for a short period of time and it is then classified on the basis of flame spread and sustained flaming.

Surface spread of flame

In the surface spread of flame test, wood-based panel products which have a density of at least 400Kg/m^3 meet the requirements of class 3 of British Standard 476 Pt7 1971. The class 3 rating enables many of the wood products to be used in buildings without special treatment, and with appropriate treatment by impregnation with a retardant during manufacture, or by surface treatment, or with a combination of these Class 1 can be achieved. Conversely, some coatings such as nitrocellulose lacquers, can downgrade a material.

Rate of heat release

The contents of a room or building frequently contribute more to a fire than the structure, and indeed, they often are the source of the primary ignition. Wood and wood-based products normally fall far below the zero rating for heat release as defined in British Standard 476 Pt 6 1968 "Fire Propagation Test for Materials", and only specialised wood products may be used in the most exacting conditions.

Fire resistance

As the title implies, fire resistance is the ability of an element to continue doing its job in a fully developed fire; it is not defined by the nature of the material used but by the combination of all the components. Test methods for the elements of building construction are given in British Standard 476 Pt 8 and ISO 834.

Because timbers char at rates which are proportionate to their density and because the charring rate is little affected by the severity of a fire, their fire resistance can be estimated and the fire resistance of simple timber elements like beams and columns can be predicted.

For commonly used softwoods with a density above 420 Kg/m^3, the rate of charring is assessed at 20 mm in 30 minutes on each exposed face, and at 25 mm for less dense species. Some hardwoods such as oak used for the construction of buildings have been awarded a 15 mm rate of charring per 30 minutes. Much fundamental work on the fire resistance of common timber structures has been undertaken by the Timber Research and Development Association, and is available in its Information Sheets "Timber Elements of Proven Fire Resistance". For proprietary forms of construction the manufacturer will usually have had the necessary tests carried out and certified.

Flame retardant treatments for wood

Flame retardant treatments for wood fall into two categories:

- impregnation with inorganic salt solutions or with leach resistant chemicals;
- surface coatings.

Impregnation with inorganic salt solutions

Dipping does not result in salt loadings which meet the relevant British Standard, hence impregnation with fire retardants requires the use of pressure treatment plant and techniques similar to those used for treatment with wood preservatives.

Inorganic salt treatments have several advantages: they are often cheaper than the materials used in surface coatings; because they penetrate the wood tissue, they are more permanent and are less likely to be removed by subsequent processing or by abrasion or other accidental damage; and fungicides and insecticides can be included with the fire retardant chemicals.

Their main disadvantages include the long process time for their application the physical size of the plant needed to treat large timbers, the relatively wasteful use of chemicals deeply embedded in the wood where they may not contribute substantially to reducing the surface spread of flame, which is essentially a surface phenomenon; possible leaching losses; possible problems with the corrosion of metal fittings; adverse effects on glues and paints; and the leaching of salts from the surface under conditions of high humidity. Corrosion problems can be reduced by the inclusion of inhibitors, and surface treatment (some paints and varnishes) can lessen or eliminate leaching.

There can be a loss of strength during the drying process which follows impregnation, because cellulose can be attacked by the inorganic salts at the temperatures involved. Losses of between 15 and 20 per cent of the bending strength have been recorded in instances where temperatures of 65°C have been used for drying. Consequently, it is wise to assume a loss of strength of 20 per cent in structural calculations.

Panel products can suffer a greater loss of strength, and the guidance of the manufacturer or processer should be sought.

Leach resistant treatments

Leach resistant treatments are applied using the same type of plant that is used for inorganic salt solutions, but the chemicals employed are monomer resins which are subsequently polymerised within the cell structure of the wood by a drying and curing cycle. Timber treated in this way does not entail the disadvantages listed above, in fact it can gain somewhat in strength and in resistance to attack by insects and fungi. Because there is no leaching problem the treated timber can be used externally or in conditions of high and fluctuating humidity; moreover, it can be painted or varnished.

Surface coatings

Surface coatings react to fire in a number of ways. Some preparations swell up, foam and char and so trap a layer of gas against the surface thus insulating it from the heat of the fire, and reducing the flow of oxygen to it; others give off inert gases which reduce flaming; and another type forms a barrier to the ingress of oxygen as well as to the escape of combustible products. Surface coatings are relatively cheap to apply; they can have a decorative function; and they are cost-effective in the sense that the maximum inhibitory effect on flaming without affecting the inherent properties of the wood or panel product.

Their disadvantages relative to impregnation treatments are their lack of a "second line of defence" against fire and their sensitivity to moisture. And, of course, their value is completely negated if they are removed during decoration and not replaced by a comparable product.

Low flame spread wood-based panels

Wood-based panels may be impregnated with flame retardants before or after manufacture to give Class 1 performance ratings under British Standard 476 part 7 1971; some have obtained Class 0 certificates under BS476 part 6 1968. Similarly it is possible to coat board surfaces or the layers near the surfaces with fire retardants during manufacture. Boards which have been impregnated during the production process may receive further surface coating after manufacture to improve their performance.

Further information

The Timber Research and Development Association is a world authority on the fire resistance of timber structures and retardant treatments, and has undertaken extensive testing and research for more than two decades. The following are some of the many leaflets currently (1990) available from the Association:

Timber Building Elements of Proven Fire Resistance	Sheet 1–11
Technology of Fire Resistant Doorsets	Sheet 1–13
Flame Retardant Treatments for Timber	Sheet 2/3–3
Low Flame Spread Wood Based Board Products	Sheet 2/3–7
Timber and Wood-based Sheet Materials in Fire	Sheet 4–11

10

Properties of the Coniferous Species (Softwoods)

The spruces

Sitka spruce dominates British softwood sawlog production. Not only is the area planted greater than that of the other coniferous species, but high rates of growth coupled with the good cylindrical form of the logs and thin bark give a yield of sawn timber per unit area that is often higher than that of other conifers. With the less commonly planted European (Norway) spruce it accounts for more than half the production of home-grown softwood.

Although there are appreciable differences in the wood properties of the two species of spruce, they are on the whole unimportant when compared with the variability within the species. After the log has been sawn it is virtually impossible to distinguish between them visually and for most marketing purposes they can be regarded as one timber.

	Bending	**Stiffness**	**Impact**	**Compression Parallel to the Grain**
	N/mm^2	N/mm^2	m	N/mm^2
European spruce (air-dry)	66	8,500	0.51	34.8
Sitka spruce (air-dry)	67	8,100	0.51	36.1

N = Newtons
Data from PRL Bull. 50

Both species are well suited for pallet boards, pitwood, boxes, packing cases, cable drums and domestic flooring. Machine graded wood is necessary for structural work such as trussed rafters and timber-frame housing. Selected timber of both species can be used for joinery —

European spruce is likely to give a better performance because of the lower incidence of spiral grain.

Spruce is a preferred material for mechanical pulping because it gives a relatively white pulp that requires little or no bleaching and the addition of relatively small proportion of chemical pulp for newsprint and other printing papers based on mechanical pulp.

Sitka spruce (*Picea sitchensis*)

When grown under British conditions the round timber may have a few narrow rings adjacent to the pith. These are the result of checked growth after planting. With modern planting, weeding and fertilising techniques there is less likelihood of a period of check after planting and rapid growth soon takes place resulting in wider rings. These become somewhat narrower in middle life and ring width continues to diminish to the end of the rotation. This growth pattern is, of course, influenced by the thinning régime adopted and by other factors such as seasonal rainfall and temperature and, in any case, it exhibits wide variability. The outer band of sapwood, which is readily visible in the fresh-felled state tends to be narrower than in most other coniferous species. Once the timber has dried out the sapwood is visibly indistinguishable from the heartwood.

The appearance of the sawn wood varies somewhat in both texture and colour — the latter can range from off-white to pale greyish-brown. Knots are normally small (ie less than 25 mm in diameter), hard and widely spaced; the distance between whorls of knots often exceeds one metre. Dead knots are less troublesome than in many other species in that they usually remain tight because the bark encasing them is thin, and the shrinkage of the knot relative to the surrounding wood tissue is small. While large knots and knot holes are seldom a problem in this species, the hard knots in a matrix of much softer wood tissue can cause damage to fast-rotating cutters resulting in a poor finish to machined wood (see page 73).

The main problem in the utilisation of Sitka spruce is undoubtedly the high incidence of spiral grain. Although this is of little consequence in its use as round timber it can be a major cause of distortion (twist), during the drying of sawn wood. For premium markets such as building timber and especially in the fabrication of trussed rafters and prefabricated components for timber frame houses, this is a disadvantage which can, of course, be overcome by grading out distorted pieces after drying.

Sitka spruce is usually regarded as a low strength timber and indeed when graded to the GS grade of British Standard 4978* it qualifies for the lowest strength class of BS 5268**, indicating that it has in general lower

* BS4978 "Timber Grades for Structural Use".

** BS5268 "Structural Use of Timber Part 2 Code of Practice for Permissible Stress Design, Materials and Workmanship".

strength properties than most other imported and home-grown softwoods. On the other hand because of its low density, it has good strength/weight ratios. For this reason Canadian Sitka spruce was a preferred material in the days of wooden aircraft manufacture.

The relatively large proportion of wide-ringed Sitka spruce failing to meet the requirement of the British Standard for joinery and the common occurrence of boards with a high percentage of juvenile wood, means that careful selection is needed if Sitka spruce is to be used for joinery or kitchen furniture. For example it is often difficult to obtain a good finish with machine tools. The poor finish results from a number of causes. Spiral grain can cause "picking" of the surface. The bands of softer early-wood tend to be pounded down rather than cut by fast rotating tools after which they recover to produce a woolly finish. As already noted hard knots in soft wood tissue can cause damage to cutters and to accentuate the tendency of the cutters to pound the softer tissue adjacent to them.

Sitka spruce is not durable and cannot be expected to last for many years when used in contact with the ground. The heartwood is impermeable to wood preservatives applied in conventional treatment plants and the sapwood can be penetrated only in the undried conditions. Effective treatment of round timber is possible using special techniques such as sap-displacement or diffusion. The former method requires the presence of a high proportion of sapwood.

European (Norway) spruce (*Picea abies*)

European spruce produces generally narrower logs, usually with a wider band of sapwood, than Sitka spruce; but once the timber has dried out the sapwood is not readily distinguished. As already noted, the sawn wood is virtually identical in appearance to Sitka spruce having a similar colour, texture, knot size and distribution. European spruce often shows a more even rate of radial growth than Sitka spruce and so the variation in ring width is somewhat less pronounced for trees grown under similar conditions.

The strength properties are so close to those of Sitka spruce that the same settings can be used on grading machines when mechanical stress grading is undertaken.

The timber is not inherently strong but has a good strength weight ratio.

In drying there is likely to be less distortion due to twist than in Sitka spruce and the shrinkage is marginally lower in both the radial and tangential planes; movement after drying is similar, both species being less troublesome than most other softwoods.

European spruce heartwood is impermeable to wood preservatives and the sapwood is only permeable before it is dried. However, because the proportion of sapwood is larger than in Sitka spruce higher loadings of preservatives can be obtained especially in the roundwood.

As with Sitka spruce, hard knots and bands of soft earlywood can result in poor finishes when the wood is machined, although there is less "picking" of the surface because of the relatively low incidence of spiral grain. Because of the radial growth pattern, a higher proportion meets the British Standard specification for Joinery.*

The pines

The pines comprise principally Scots pine, Corsican pine and lodgepole pine. The properties of these species are dissimilar and for many purposes it is prudent to market them separately. Small quantities of other pines, such as maritime pine (*P. pinaster*), radiata pine (*P. radiata*), Austrian pine (*P. nigra var. austriaca*) and yellow (Weymouth) pine (*P. strobus*) are grown in Britain, but neither individually nor collectively do they play any significant part in the British woodland produce market.

Scots pine (*Pinus sylvestris*)

Although a native tree, the timber is more likely to be encountered in its imported form as redwood from the Baltic area or the Soviet Union. Because for roughly two centuries it has been the most familiar all-purpose softwood, it is regarded by the timber industry as the "yardstick" against which other coniferous timbers are assessed. In Britain the tree does not grow to a great height and this feature — coupled sometimes with crown damage resulting from attack by insects and squirrels — tends to produce shorter logs of sawmill quality than the spruces and Douglas fir; poor stem form together with a thick bark also contributes to a lower yield of sawn timber.

The occurrence of logs of veneer quality and of poles meeting the requirements of British Standard 1990 is comparatively small, but where trees of this standard exist they are worth marketing separately.

The strength of the timber is loosely correlated with the ratio of early (spring) wood to the stronger late (summer) wood in the annual ring; in practical terms this means that very fast grown timber with only a few rings per 25 mm is the weakest followed by very slow grown timber with up to 20 rings per 25 mm. Much British grown sawn timber has strength properties equal to those of *Pinus sylvestris* from southern Sweden. This means that it is in general weaker than that from Poland and the former Baltic States but stronger than that from the northern areas of Sweden, Finland and the Soviet Union.

* BS1186 "The Quality of Timber and Workmanship in Joinery".

While the colour of Scots pine varies considerably, the creamy-white sapwood is quite distinct from the pale orange-brown heartwood. The growth rings are also fairly prominent because the denser latewood is somewhat darker than the earlywood.

The orange-brown knots are frequently loose because they may be encased in a thick ring of bark from which they shrink on drying.

The timber works well with machine and hand tools but with a slight tendency to split around nails in dry timber. Although it is not durable it absorbs preservatives well. The sapwood is susceptible to attack by blue stain fungi, especially during the humid late summer and autumn months.

Although the timber offers no significant problems in drying in some situations the dried wood exhibits appreciable movement in changing conditions of atmospheric humidity. This can cause undesirable gaps to appear in joints, and can create a decay hazard in external joinery where no preservative treatment has been applied.

Corsican pine (*Pinus nigra* var. *maritima*)

The roundwood of Corsican pine exhibits good cylindrical form, but with large widely-spaced whorls of knots (the internodes sometimes exceed one metre in length), together with a much larger proportion of sapwood than in other conifers. The fine stem form usually ensures that the yields of poles meeting the requirements of BS1990* is high and that the sawlogs are straight.

On the other hand the knot whorls can result in planes of weakness in both bending and impact loading. They do not affect the compressive strength parallel to the grain significantly.

The extensive sapwood facilitates deep penetration by and good retention of wood preservatives; consequently Corsican pine has sometimes been favoured for piles and jetty timbers where the risk of attack by marine borers is known to be great.

The long internodes of comparatively straight grained wood, which enable lengthy strands with adequate tensile strength to be cut, have made it the preferred species for the manufacture of wood-wool and wood-wool slabs.

The timber air dries relatively slowly and during the late autumn and winter months it has a tendency to pick up moisture from the atmosphere. Moreover, in partially dried wood there is a risk of attack by blue stain fungi. Kiln drying should be undertaken if the appearance of the finished product is important. After drying it is less prone to movement in conditions of changing humidity than Scots pine.

* BS1990:1984 "Wood poles for overhead power and telecommunication lines" [Note, Corsican pine is no longer accepted by British Telecom]

The sawn wood is similar in appearance to Scots pine, but it usually has wider growth rings and a coarser texture. It works and finishes well.

Lodgepole pine (*Pinus contorta*)

Lodgepole pine exhibits considerable variability in stem form. Inland provenances often produce lightly branched trees which yield straight logs of comparatively slow growth. The coastal provenances, which grow faster in their early years, tend to have rather bowed stems and heavy branches. The species is a relative newcomer to British forestry and for this reason much of the published information on its wood properties is based on test material from the Irish Republic.

The timber in many respects resembles that of Scots pine. The appearance of the sawn wood is almost identical but its strength properties are somewhat lower. No development work has been undertaken on the machine stress grading of home-grown lodgepole pine and so it should be avoided for load-bearing purposes in building unless designs have been specially prepared and approved.

Due to the low contrast between the earlywood and latewood and the relatively soft texture of the knots, it works better than any other common British softwood. Quite wide ringed timber gives a near-perfect finish when machined.

It shrinks less than Scots pine and is somewhat more stable in changing humidities, hence it gives a much better performance when used for furniture and joinery even though the rate of radial growth can be greater than that required by the British Standard specification for joinery.

The larches (*Larix decidua, L. Kaempferi* and *L. x eurolepis*)

Larch is the strongest of the plantation-grown softwoods. An extensive assessment of the timber properties of European and Japanese larch, together with their hybrid, Dunkeld larch, by the former Forest Products Research Laboratory in the 1960s demonstrated that the wood properties are so close compared with the variability within the species that, with the exception where appearance is important, they can be marketed together as one timber. Initially Japanese larch was considered to be inferior, but this was the result of comparisons of small thinnings of this species containing a high proportion of juvenile wood with mature European larch (see also page 89). In fact, the strongest larch tested so far is Japanese larch from the Forest of Dean.

The superior natural durability and good strength properties of larch make it the preferred timber for estate work; treatment with a preservative is recommended where it is to be used in contact with the ground for more than a decade (see page 112).

Although there are some notable examples of larch interior joinery, it is apt to split when nailed unless the holes are pre-bored; there can be problems with resin exudation; and unless careful selection is undertaken there may be a high incidence of dead knots which tend to fall out during drying. It is occasionally used for veneers, where the warm golden brown tones of European larch in particular give a pleasant ornamental effect.

Selected poles have been used for telecommunications and for electricity supply, but the yield per hectare of acceptable poles meeting the rigid straightness criterion is low.

The highest grade of sawn larch is traditionally used for fishing boat-skins and to a lesser extent for the production of certain types of vat. For these uses the durability and the relatively small knot size are advantageous. Boat-skin larch planking may be sawn parallel to the bark to produce long lengths from the outer parts of slow-grown knot-free logs.

Other uses include fencing, especially garden fence panels (see page 86), pitwood, and some types of pulpwood. It is less favoured for particle board manufacture because it increases the weight of the panels and darkens their colour without providing any notable advantage. When used for fuel on open fires it is likely to "spit".

Douglas fir (*Pseudotsuga menziesii*)

Mainly as a result of the variation in its strength properties, Douglas fir has been an enigma to the users of British softwood. Early tests at the former Forest Products Research Laboratory during the 1950s based on sample logs taken from a number of private estates indicated that it had fairly high strength in all facets other than splitting and shear, being only marginally less strong than the Canadian equivalent and roughly on a par with European larch. Similarly, an assessment of the compressive strength of 180 pit props of various dimensions confirmed that it was significantly stronger than the pines and spruces. In fact, Douglas fir was the only species with a strength consistently exceeding one long ton per square inch [15.5MN/m^2] in the undried condition (European larch was not included in this project). As a result Douglas fir was given initially a high rating for structural work. Subsequently, with the advent of mechanical grading, yields of the stronger grades were found to be unacceptably low, with the result that it was "demoted" to strength ratings roughly equivalent to those assigned to European spruce.

Douglas fir is available in long lengths from trees of large diameter and is, therefore, used for industrial and other large buildings (Figs 10.1 and 10.2). Buildings that employ Douglas fir for the main structural elements include the terminal buildings at Plymouth Airport; the village hall at Bratton Clovelly; the Visitor Centre at Westonbirt Arboretum near Tetbury; and the timber demonstration building at the Farm Buildings Information Centre near Kenilworth.

An off-white sapwood contrasts with a salmon pink heartwood; the latter is disliked for some purposes, but it ages into a pleasant mellow brown which is of value when used for internal joinery. The wood is also characterised by a marked contrast in density between the earlywood and the much harder latewood, although there is less colour contrast within the growth ring than is observed in larch.

Large knots may occur relatively frequently. Careful selection is therefore needed when Douglas fir is used for joinery, load-bearing structures or flooring; but given due compliance with the appropriate British Standard, it will give good service in all types of building work. For heavy duty flooring it may be necessary to specify or select quarter (rift) sawn wood to avoid the possibility of "shelling out" of growth rings.

It works and finishes well, although inadequate sharpening and setting of tools can cause splintery surfaces. It is also apt to split on nailing, especially near the ends of boards, consequently the pre-boring of holes is advisable when it is used for exacting purposes.

Douglas fir is only moderately durable and preservative treatment under pressure is essential for lasting service out of doors (it is resistant to impregnation and it used to be common practice to pass timbers (e.g. sleepers) of this species through an incising machine prior to pressure treatment in an attempt to improve the retention of preservative).

Fig 10.1 Wide span portal frame structure — village hall Bratton Clovelly, Devon. Courtesy Michael O'Connor

It can be regarded as a good general purpose softwood suitable for telegraph and power transmission poles, round and sawn fencing, all building work including pole-barns, packaging and materials handling.

Fig 10.2 Terminal building Plymouth Airport, constructed in 1980 using locally grown Douglas fir for the main structure. Courtesy Michael O'Connor

Some less commonly planted conifers

Austrian pine (*Pinus nigra var. nigra*)

Austrian pine is an alpine variety of the same species as Corsican pine, but it is reputed to yield timber with larger knots when grown under plantation conditions. Grade for grade it can be used for the same purposes.

Port Orford cedar commonly called Lawson cypress (*Chamaecyparis lawsoniana*)

An almost featureless even-textured creamish-white wood with a persistent spicy scent and displaying no pronounced difference between earlywood and latewood, it has moderate strength properties. It is fairly durable making it well suited for fencing and estate work. While its aroma makes it unsuitable for use in contact with foods, it has given it a reputation for being moth-repellent and for this reason it is used in its native north America for lining wardrobes and chests-of-drawers. The foliage, which can be extremely varied on account of the number of ornamental cultivars available, is valued for wreath-making and floristry.

There has been limited testing of the botanically related hybrid Leyland cypress (*Cupressocyparis leylandii*). While the wood properties appear to be similar to those of Port Orford cedar, it is rather more durable and it might well prove to be the most durable of all home-grown timbers.

Silver firs (*Abies spp*)

Firs, particularly the European silver fir (*Abies alba*), grand fir (*Abies grandis*) and noble fir (*Abies procera*), produce an off-white to pale brown timber superficially resembling spruce,* and with comparable strength properties, but capable of giving a better, less woolly finish. They are also more amenable to treatment with wood preservatives. They have good nailing properties and are well-suited for pallet boards; selected wood has been used for kitchen furniture and joinery. Silver firs should be avoided for load-bearing structures as there is no provision for their use under the Building Regulations. The foliage is readily marketed for floristry, and the tops and saplings make the best Christmas trees with little or no loss of needles.

Western hemlock (*Tsuga heterophylla*)

The timber is an almost featureless pale brown colour with no marked difference between heartwood and sapwood. It is somewhat weaker than western hemlock from Canada,** with strength properties falling between those of Scots pine and spruce, except that it has lower values for stiffness. It is best regarded as a non-durable general purpose softwood with no outstanding properties other than a reputation for gluing well. The foliage is used in floristry.

Western red cedar*** (*Thuja plicata*)

A red-brown wood of good natural durability. Its strength properties — which are approximately ten per cent lower than those of spruce — are

* Consignments of "Whitewood" from central Europe frequently contain both European spruce and silver fir.

** Consignments from Canada often mix hemlock with spruce and balsam fir, which are marketed as "Hem-spruce-fir".

*** The appellation "cedar" embraces a number of dissimilar and botanically unrelated species which have in common a yellowish to red-brown colour and a fragrant or spicy odour. They include the true cedars (*Cedrus atlantica, C. libani* of biblical fame, and *C. deodora*), all of which are commonly planted in parklands; the Japanese cedar (*Cryptomeria japonica*); Port Orford cedar (*Chamaecyparis lawsoniana*) popularly called Lawson cypress; the pencil cedars (*Juniperus* spp); and the cigar box cedar (*Cedrela mexicana*) which is a hardwood imported from central America for guitar making.

comparable with wood of the same species imported from Canada. Because of its low density the timber has excellent strength/weight ratios. Although it has some ornamental value, its natural durability is the main factor in its utilisation for such purposes as external joinery, light-weight gates, seed-boxes and greenhouses. In damp locations it can be corrosive to steel, and it is, therefore, important to use galvanised fastenings. Its performance for roof shingles in the UK has been disappointing. The foliage is in regular use in floristry, especially for wreath making.

Yew (*Taxus baccata*)

Although it is not grown commercially yew is the strongest, heaviest, hardest and one of the most durable softwoods found in the British Isles. A narrow band of white sapwood is readily distinguished from a much darker heartwood. The annual rings are pronounced. The earlywood is cinnamon brown. It contrasts pleasantly with the "plain chocolate" coloured bands of latewood. Yew played an important role in our military history, being the traditional material for the long bow. Nowadays it is prized for its decorative properties, particularly for veneers, furniture and for ornamental turnery (Fig 10.3); even quite short logs can be used for turning into such items as drawer handles, door knobs, egg cups and pepper mills.

Sequoia (*Sequoia sempervirens*)

A few plantations of sequoia exist in milder areas of Britain. It is a soft low density timber in which a highly durable ruddy brown heartwood contrasts with a creamish white sapwood. While it has relatively low strength properties, it has some ornamental value and has been used for veneer manufacture. However, it is best regarded as an estate timber, well-suited for fencing and light-weight gates (except in those situations where there is an abnormal risk of abrasion or impact).

The wood of the popular parkland tree Wellingtonia (*Sequoiadendron giganteum*) [formerly *Sequoia gigantea*] has timber of similar properties and appearance, but on account of its conical stem form it produces somewhat wasteful sawlogs.

The foliage of both species is useful for wreath-making and floristry.

The identification of home-grown softwoods

Even with the use of a microscope it is often quite impossible to determine the species of a coniferous timber; for example the anatomical structures of European and Sitka spruce are indistinguishable. On the other hand it is usually feasible to identify the genus. The following notes have been

Fig 10.3 Table lamps in English yew. Light tones are sapwood. Heartwood darkens in light. Pin knots are regarded as a decorative feature. Turned by R E Crowther.
Photo, courtesy Forestry Commission

prepared on the assumption that no equipment other than a hand lens is available.

Although the separation of the genera remains a difficult task, the gross features together with the revelations of the hand lens can provide useful pointers. For example, if blue stain is present in a softwood which is known to be home-grown, the field is narrowed to the genus *Pinus* and Douglas fir. Of course the sapwood does not have to be infected by blue

Fig 10.4 Sitka spruce × 5. End-grain of Sitka spruce (*Picea sitchensis*) a lower density softwood showing the relatively small proportion of latewood which displays less contrast with the earlywood than most British conifers. Note the fine fairly uniform rays, and the sparse, apparently randomly distributed, narrow, vertical resin canals.

Fig 10.5 European spruce × 4 Transverse section of European spruce (*Picea abies*) showing the smaller thicker-walled tracheids in the latewood. Note the paucity of the vertical resin canals (in this sample near but not within the latewood), and the extreme narrowness of the rays.

Fig 10.6 Scots pine × 5
End-grain of Scots Pine (*Pinus sylvestris*), a medium density softwood showing a marked contrast between the latewood (with smaller thicker walled tracheids) and the earlywood, and also narrow fairly uniform rays. Note that the majority of the vertical resin canals concentrated in or near the latewood.

stain fungi, in which case an examination of the end grain with the hand lens is necessary.

The use of gross features

Annual rings A pronounced contrast between early wood and late wood is observed in both larch and Douglas fir. The former also exhibits a marked colour difference (ie darker late wood) giving a "tiger stripe" effect.

Knot size Knots exceeding 25 mm in diameter are much more common in Scots pine, Corsican pine, lodgepole pine and Douglas fir than in other species.

Dead knots Dead knots encased in relatively thin bark in the spruces tend to be tighter than in other species.

Bark retained on waney edges Thick red bark usually indicates Scots pine.

Finish on rough-sawn tangential surfaces A relatively smooth surface may be indicative of the pines, a woolly but flat surface may be expected in the spruces, while a "relief" surface resulting from the contrast in texture between the earlywood and latewood is often observed in the larches and Douglas fir.

Hand lens examination of the annual rings on the end-grain

The spruces display a low contrast between earlywood and latewood, with no obvious distribution pattern in the axial (longitudinal) resin canals.

The larches and Douglas fir show a sharp contrast between the earlywood and latewood; the resin canals are barely perceptible and have no obvious distribution pattern.

The pines have their resin canals concentrated mainly in the latewood.

The performance of home-grown softwoods

The following "league tables" list the British coniferous species in order of performance. The strength ratings refer to air-dry timbers.

Nominal specific gravity*	Compressive strength**	Bending strength
European larch	Douglas fir	European larch
Scots pine	Scots pine	Douglas fir
Douglas fir	European larch	Scots pine
Japanese larch	Corsican pine	Japanese larch
Corsican pine	Japanese larch	Corsican pine
Lodgepole pine	Dunkeld larch	Lodgepole pine
Dunkeld larch	Lodgepole pine	Dunkeld larch
European spruce	Sitka spruce	Sitka spruce
Sitka spruce	European spruce	European spruce

Stiffness	**Impact strength**	**Shear strength**
Douglas fir	European larch	Scots pine
Scots pine	Scots pine	European larch
European larch	Douglas fir	Lodgepole pine
Corsican pine	Japanese larch	Japanese larch
Dunkeld larch	Dunkeld larch	Dunkeld larch
European spruce	Lodgepole pine	Douglas fir
Japanese larch	Corsican pine	Corsican pine
Lodgepole pine, Sitka spruce	European spruce, Sitka spruce	European spruce
		Sitka spruce

The information in the above tables is from PRL Bulletin 50 and relates to small clear specimens only. While it is useful for drawing comparisons, in practice the incidence of defects has to be taken into account (see chapter 5)

* The nominal specific gravity is the oven dry weight divided by the air-dry volume.
** Compressive strength parallel to the grain. Compressive strength perpendicular to the grain is of less importance.

11

Properties of the Broadleaf Species (Hardwoods)

The diversity of site types and species, coupled with the diversity of sylvicultural systems (eg high forest; coppice with standards; simple coppice) have made it difficult for individual woodworking industries to rely wholly on home-grown hardwoods for a long-term supply of wood of consistent quality. There have, of course, been exceptions — some turneries have succeeded in getting all their roundwood requirements over long periods; beech grown in the Chilterns supplied wood for chair parts for many decades; and willow plantations in East Anglia have supplied the cricket bat industry continuously in that region. But most hardwood-based woodworking enterprises have relied to a greater or lesser extent on imports to supplement home supplies; others have been based entirely on imports of temperate or tropical hardwoods. Even the most commonly occurring British broadleaved timbers — oak and beech — have been supplemented by the abundant supplies from countries such as Czechoslavakia, France, Romania, the United States and Yugoslavia (see also Chapter 7).

It will be noted that this chapter, unlike the previous one, contains no section on identification. This has been omitted because the situation is much more complex. Unlike softwoods, the wood structure of species within a genus (eg *Quercus*) is often very different. Those undertaking the identification of hardwood species are referred to "An atlas of end grain micrographs" For. Prod. Res. Bull. 26, and "Identification of hardwoods: A lens key" For. Prod. Res. Bull. 25. Published by HMSO. Both should be obtainable from central libraries.

Alder (mainly *Alnus glutinosa*, occasionally *A. cordata*, *A. incana* and *A. rubra*)

A dull brown-pink, light weight, even-textured wood with low strength properties and with no outstanding features other than ease of working and nailing. It has been a traditional material for clogs, hat-blocks and artificial limbs where its low density is advantageous. Otherwise it is regarded as a softwood substitute for use in non-structural building, or as a non-ornamental turnery wood. Although it is non-durable it takes preservatives readily and if treated can be used for estate work and fencing.

Ash (*Fraxinus excelsior*)
Ash is one of the most valuable of native timbers coupling good strength properties with ease of working. The off-white to yellowish-white wood contains frequent bands of relatively large pores giving it a coarse appearance, none-the-less it works well taking a smooth finish. It is among the strongest of home-grown hardwoods with a justifiable reputation for resistance to impact loading; hence it is the preferred species for tool and implement handles; for vehicle construction and repair, eg in the building and restoration of horse-drawn carriages; for the frames of estate cars and caravans; and for sports goods. Coppice-grown wood is sometimes specified for these uses.

Careful selection is needed to ensure strong wood, and it is usual to specify a number of annual rings per 25 mm of radius (the best material has between 4 and 10 rings per 25 mm). If it is grown too slowly the proportion of pores relative to the fibres in the wood tissue increases, resulting in a marked loss of strength. The slower grown material is, however, well-suited for ornamental use in furniture manufacture and selected logs are sometimes acquired for the production of decorative veneers. It is not durable and where it is used for estate work or for ladder rungs, pressure treatment with a preservative is necessary (round fence posts of this species treated by the open tank process have given disappointing results).

Beech (*Fagus sylvatica*)
Long periods of selective cutting involving the removal of the best stems, together with the use of seed from inferior parent trees has resulted in many stands of beech having a poor log form with a low yield of good quality sawn timber.

It is an even-textured, coffee-coloured,* diffuse-porous wood with no pronounced difference between sapwood and heartwood and with good all-round strength properties. While the rays are large enough to be visible to the unaided eye, there is hardly any contrast with the rest of the wood tissue and consequently they do not give an appreciable ornamental effect.

Beech works, nails and screws well. It has been the traditional material for making many types of tool such as jack planes, mallets and file handles. The higher grades of sawn and turned wood are in continuous demand for furniture manufacture where they may be used in conjunction with more ornamental species; for example a Windsor chair normally has a beech back and legs with a more decorative elm seat. Similarly an oak cabinet may well have the "out-of-sight" parts such as drawer backs and sides made of beech. For these and like uses it can be stained to match the

* If beech is kiln dried at high temperatures it acquires a not unattractive pink tinge.

species which it complements. Beech on its own is widely used for school furniture. It is a traditional material for steam bent work such as that used in the manufacture of occasional chairs. Beech plywood is a major product in eastern European countries.

It can make an attractive domestic floor (parquet, block or strip), especially if the lower grades are used to give it "character". It is usually avoided for joinery in favour of more stable or lower density timbers. Other indoor uses include toys, models, and "trees" for boots and shoes.

While it is not resistant to decay, it is permeable to wood preservatives and if treated under pressure it is well suited for estate work including fencing (in some European countries creosoted beech is the preferred material for railway sleepers, but sleepers made from temperate hardwoods have not been favoured in Britain).

Birch (*Betula pubescens* and *B. pendula*)

While birch is one of the strongest home-grown woods, its uses are limited by its inability to attain wide diameters under British conditions and the poor form of the available logs; boards exceeding 300 mm in width are rare. There is nowhere near enough birch of the required quality to sustain a birch plywood industry like that found in Finland where logs in the relatively small diameters achieved by birch are used extensively.

The diffuse-porous, off-white to pale pinkish-brown wood is almost featureless. There is no obvious difference between heartwood and sapwood, there is no ray figure, and the annual rings are apparent only as faint bands of terminal fibres. It works and stains well. It can be used in much the same way as beech for furniture manufacture, toy making, tool handles, flooring and non-ornamental turnery. It is the traditional material for clothes pegs. Although it is not resistant to decay, it takes preservatives well and, if treated, can be used for fencing and other estatework.

Cherry (*Prunus spp.* mainly the European [wild] cherry *P. avium*)

Cherry is an even-textured diffuse-porous wood in which the pale sapwood is quite distinct from the medium-brown heartwood. The latter is not unlike some tropical hardwoods in appearance, but sometimes displays a greenish tinge. It has good strength properties, being superior to oak. The best quality timber can be used for furniture production, panelling and cabinet making, while lower grades are suitable for estate work. As it is resistant to penetration by wood preservatives it should be avoided for service in situations in contact with the ground. Cherry is a good turnery timber (Fig 11.1).

Sweet Chestnut (*Castanea sativa*)

Sweet chestnut has the desirable attributes of ease of working, stability in changing conditions of humidity, good natural durability, attractive appearance (a figure resulting from wavy grain is sometimes present) and moderately good strength properties. The frequent incidence of shake in larger trees, with a consequent reduction in yield, is the only serious disadvantage in an otherwise valuable timber.

The light brown heartwood resembles that of oak, but its rays are much smaller, hence it lacks the silver grain figure characteristic of oak. Its strength properties are slightly lower. The sapwood is a much paler creamy brown and it forms a much smaller proportion of the log than in oak.

While it is a good timber for furniture production and internal joinery, it is a near-perfect wood for outside work being suitable for such diverse uses as the glazing bars and mullions of windows to fencing (where it can be used in contact with the ground without preservative treatment). It has a tendency to split in nailing unless the hole is pre-bored. For out-door work galvanised nails and fittings should be used to preclude the possibilities of unsightly stains and corrosion.

Coppice grown chestnut is the preferred species for hop poles, bean rods, walking sticks and umbrella handles and cleft chestnut fencing pales (see chapter 6).

The timber of the unrelated horse chestnut (*Aesculus spp.*) is completely different. It is a white diffuse-porous wood, which is virtually all sapwood with moderately low strength properties (it has poor values for stiffness). As it is available in wide boards it is particularly suitable for shelving in dry places where the decay hazard is low.

Elm (*Ulmus spp*)

The sapwood which is yellowish-white is distinct from the golden brown or reddish brown heartwood. Wild or cross grain occurs frequently, but even in straight-grained wood the strength properties are moderately low being roughly equivalent to those of spruce. Careful selection is needed if it is to be used in any load-bearing situation; it should be avoided for such purposes as dowels.

On the other hand the figure produced by the "undisciplined" grain, together with the zig-zag appearance of the tangential surfaces (resulting from the bands of smaller pores) and the delightful ray figure on the radial surfaces, make it arguably the most handsome of home-grown hardwoods.

Prior to the ravages of the Dutch elm disease it was available in good widths making it a preferred timber for coffin boards, garden furniture and chair seats (especially Windsor chairs). For these purposes careful kiln drying is essential to preclude the possibility of warping in service. It is also used for civil engineering; pallet blocks; sea defences; port installations; and weather boarding. For these purposes treatment with a wood preservative is desirable (it is permeable).

Wych elm (*Ulmus glabra*) resembles the other species superficially, but it has somewhat better strength properties, generally superior to those of oak, and it usually has a straighter grain. The finished wood has a lustrous green tinge. In addition to the above uses it has been in demand for boat building, and is used by wheelwrights.

Hornbeam (*Carpinus betulus*)

A diffuse porous white timber displaying no obvious difference between heartwood and sapwood. It is the strongest, hardest and heaviest British timber, but due to its limited distribution and poor stem form, good quality sawlogs are relatively rare. It was formerly used for machinery parts and other objects now made from metal or from even harder tropical hardwoods. It makes an excellent industrial floor or it can be stained for domestic flooring. Although it is not durable it is readily treated with preservatives and thus can be used for practically all estate work especially in situations where a high resistance to abrasion is required.

Lime (*Tilia spp.* mainly *T. vulgaris*)

A diffuse-porous even textured pale brown wood with moderate strength properties. Lime is the preferred species for wood carving, being easy to work in all three planes; it is also favoured for the internal parts of beehives on account of its freedom from taints. Other uses include kitchen implements, brushware, non-ornamental turnery and as a softwood substitute. It accepts preservatives readily and can thus be used for estate work.

An infusion of dried lime flowers is a popular beverage in some European countries. It resembles weak tea and is often served with lemon.

Oak (*Quercus robor* and *Q. petraea*)

The ready availability of oak in Britain has caused it to become the yardstick against which other hardwoods are assessed. Certainly its good natural durability, its handsome appearance and its resistance to abrasion have been put to good use since the stone age; but hardness apart it does not have any outstanding strength properties, and its strength weight ratio can only be regarded as moderate.

The off-white sapwood is considerably paler than the light brown heartwood, and it can form a high proportion of both round and sawn oak (for this reason care is required in the selection of posts to be used without preservative treatment).

The uses for oak are very dependent on the grade, for a wide range of qualities may be encountered. As already noted on page 194 selected logs

Fig 11.1 Bowl in cherry. Pepper mill in oak: dark streak is a natural stain in the wood from a wound. Turned by R. E. Crowther
Photo, courtesy Forestry Commission

are sought for the manufacture of ornamental veneers, and are sometimes exported for this purpose.

The higher grades of sawn timber are used for coffin boards, shop fitting, panelling, furniture, joinery, flooring, the repair of historic buildings, the frames of mock Tudor houses, the ribs and frames of wooden boats, ladder rungs and turnery (Fig 11.1). For many of these uses, full advantage may be taken of the highly ornamental silver grain.* Smaller sections of clear quarter sawn oak were formerly in brisk demand for tight cooperage, where its relatively large shrinkage and swelling was advantageous, but this industry has virtually disappeared.

Lower grades of oak are useful for industrial flooring, fencing, estate work and civil engineering. They were formerly in demand for the construction and repair of railway waggons.

British-grown red oaks (*Q. spp.* especially *Q. rubra*) and the Turkey oak (*Q. cerris*) may be available in small quantities. While they are stronger than the native oaks, they are less durable, and having no tyloses,** they are permeable to fluids, and are thus unsuitable for tight cooperage. They are permeable to wood preservatives and can be stained readily. Quantities of red oak are imported from the United States for the production of Tudor/Jacobean styles of furniture.

Poplar (*Populus spp.* incorporating a wide range of hybrids, varieties and cultivars)

Although a light weight timber with relatively low strength in all properties other than stiffness, poplar has a number of unique advantages. The colour is usually off-white with no clear difference between heartwood and sapwood, but it can be grey, pinkish or pale brown. However, its clean appearance coupled with the absence of any taints makes it favoured for use in contact with food, and it is used in the production of trug baskets, lettuce and broccoli crates.

It was formerly in demand for fruit punnets and chip baskets, but in recent decades it has been ousted by plastic or paper containers. For the production of crates short logs are peeled without pre-heating (poplar is one of the few timbers which can be peeled cold) into continuous veneers, which are then guillotined into slats.

Poplar was also used extensively for the manufacture of matches, but the post-war years have seen the closure of the three remaining British match splint factories in favour of imported splints from Canada.

Poplar is difficult to ignite as it tends to smoulder rather than burst into flames and hence it is the preferred material for oast house floors, and for other situations with a fire hazard. In the manufacture of matches the flammability is regulated by impregnation with a controlled amount of hydrocarbons.

When subjected to abrasion it bruises rather than becoming splintery, and for this reason it is favoured for waggon bottoms; but it should not be

* The silver grain figure is produced by the large rays. It is apparent on any surface, but it becomes most prominent when the log is quarter sawn to present the optimum radial plane. Some users maintain that the best ornamental effect is given when the wood is cut at a shallow angle to the radius.

** See page 184

used for cattle trucks or horse boxes where the risk of decay is somewhat greater.

In fixing poplar it is important to use flat-top nails or staples, because panel pins and similar types of narrow nail tend to "pull through". For the same reason it should be avoided for pallet boards.

It is not durable and is resistant to preservative treatment and so is not well suited for estate work.

Sycamore (*Acer pseudoplatanus*)

A most useful wood with moderately good strength properties (roughly equivalent to those of oak), good stability and a pure white colour. While it works easily giving a better finish than most other species, it is sometimes necessary to pre-bore nail holes to avoid splitting.

Logs displaying wavy "fiddle back" grain are in keen demand for the manufacture of ornamental veneers, and for the making of the backs and necks of the violin family of musical instruments.

It turns well giving a superior finish to turned beech, and it was formerly used extensively for textile woodware — especially for bobbins for the retail marketing of thread but for this use it has now been replaced by plastics.

Because of its clean colour and because it is devoid of any taints, it is the preferred timber for use in contact with food, and is the traditional material for kitchen and dairy implements, draining boards and chopping blocks. It is also suitable for furniture, joinery and toy making.

It readily accepts preservative treatment and can be used for fencing and other estate purposes; without treatment it is not durable.

Less common broadleaf species

European [London] Plane* (*Platanus acerifolia*)

The sawn timber, known as lacewood, has a highly ornamental figure resulting from the large reddish brown rays in a matrix of much paler yellow brown tissue. It is used for veneers, panelling, marquetry, inlay work and other decorative purposes. It is one of the best turnery woods.

Eucalyptus (*Eucalyptus spp.* especially *E. gunii*)

Eucalyptus grown under British conditions is unlikely to produce marketable sawlogs, but it is a high volume producer with good possibilities for coppicing, and is often considered as a possible source of pulpwood or for energy plantations. The grey-green foliage is used in floristry.

* In Scotland sycamore is commonly called "plane".

Holly (*Ilex aquifolium*)
A pure white strong wood which is prone to distortion during drying. It is used for turnery, inlay work and marquetry. It takes stain evenly, and when dyed black has been used as a substitute for ebony (chess boards have been made using holly for both the black and white squares).

Laburnum (*Laburnum spp.* mainly *L. anagyroides*)
A lustrous, durable, dark brown heartwood contrasts with a narrow band of light yellow sapwood. The tangential bands of paler parenchyma in the heartwood gives an ornamental figure; but the uses of laburnum for decorative purposes are limited both by the scarcity of supplies and the narrowness of the logs. It has been used for veneers and inlay work. It is an excellent turnery wood (Fig 11.2), suited especially for the production of wood wind instruments (it was traditionally used for bagpipe chanters).

Nothofagus [Southern beeches] (Principally *N. obliqua* British Standard name coigue, and *N. procera* British Standard name rauli)
Nothofagus has an almost featureless pinkish-brown to coffee-coloured heartwood with a somewhat paler sapwood. Although it is not attractive enough to be used for ornamental purposes, selected wood can make a pleasing domestic floor. Both species dry slowly with a marked tendency for splits and surface checks to develop. They machine satisfactorily, but as they have no outstanding strength properties, they are best regarded as general purpose hardwoods for uses such as pallet blocks.

Robinia (*Robinia pseudoacacia*)
An attractive yellow-brown timber. It has many properties similar to ash, and so it can be used for much the same purposes. It has, however, good natural durability and thus can be used out-of-doors without preservative treatment.

Walnut (*Juglans regia*)
Walnut logs for veneer manufacture, for high grade furniture, vehicle dashboards and gun stocks are likely to command higher prices than any other British timber. In addition to its ornamental value, it is strong, stable and durable.

Willow (*Salix spp*)
Willow has similar properties to poplar, but is marginally weaker; it is used for similar purposes, especially in the production of trug baskets. *S. alba* var. *coerulea* is cultivated, mainly in East Anglia, for the blades of cricket bats. In some area of southern England, notably Somerset and East Anglia the branches of coppice willows are collected annually for the manufacture of woven baskets; the branches are boiled after harvesting to facilitate the bark removal.

Fruit trees

Timber from both wild and cultivated fruit trees (mainly *Malus spp* — apple and *Pyrus communis* — pear) is sometimes available as sawn wood. The pinkish-brown to fawn heartwood (the sapwood is paler) is even-textured, hard and heavy, but with good stability and resistance to abrasion, although some varieties have a reputation for brittleness. It has been used for printers blocks; turnery; marquetry and in-lay work; drawing instruments (T-squares and set squares); and mallet heads.

Fig 11.2 Knitting needle box in laburnum. Lighter tones are sapwood. The heartwood darkens to a rich brown with exposure to light. Turned by R. E. Crowther. Photo courtesy Forestry Commission.

Sawn mining timber

Virtually every species of hardwood mentioned above has been used as sawn mining timber, a now diminishing market that once absorbed very large quantities of British hardwoods (see page 103).

The performance of home-grown hardwoods

The tables below list British hardwoods in order of magnitude of specific gravity and of performance strength). The strength properties refer to air-dry timbers.

Nominal Specific Gravity*	Compressive Strength**	Bending Strength
Hornbeam	Hornbeam	Birch
Beech	Birch	Hornbeam
Oak	Beech	Beech
Ash/birch	Cherry	Ash
Cherry	Ash	Cherry
Wych elm	Oak	Wych elm
Sycamore	Wych elm	Sycamore
Lime	Sycamore	Oak
Sweet chestnut	Lime	Lime
Elm	Sweet chestnut	Alder
Alder	Alder	Sweet chestnut
Poplar/willow	Poplar	Poplar
	Elm	Elm
	Willow	Willow

Stiffness	Impact	Hardness	Shear
Birch	Hornbeam	Hornbeam	Hornbeam
Beech	Beech	Beech	Sycamore
Hornbeam	Ash	Ash	Ash
Ash	Cherry	Cherry	Cherry
Lime	Wych elm	Oak	Birch
Wych elm	Birch	Birch	Beech
Cherry	Oak	Sycamore	Oak
Oak	Sycamore	Wych elm	Wych elm
Sycamore	Lime	Elm	Elm
Alder	Willow	Lime	Alder
Poplar	Alder	Sweet chestnut	Lime
Sweet chestnut	Elm	Alder	Sweet chestnut
Elm	Sweet chestnut	Willow	Poplar
Willow	Poplar	Poplar	Willow

Data from PRL Bull. 50

* Nominal specific gravity is the oven-dry weight divided by the air-dry volume.

** Compression parallel to the grain.

12

Residues and Produce other than Timber

In traditional tree harvesting operations the stem is all that is removed from the forest. It accounts for less than half the volume of the whole tree including roots and crown. Stumps and roots account for roughly 20 per cent; tops and branches amount to about 15 per cent; bark is between 7 and 20 per cent; and harvesting losses, mainly sawdust, account for about 5 per cent (Fig 12.1).

Stumps and roots

The removal of stumps and roots is sometimes required on pathological grounds, particularly for the control of the parasitic fungus *Heterobasidion annosus* which readily colonises stumps, from which it is able to attack and destroy a newly planted crop. In East Anglia the control of this pest is effected by bulldozing the stumps and roots into heaps. After they have dried out, they are burned.

While stumps and roots are, of course, unsuitable for sawmilling, it is technically possible to use them for pulp and board manufacture, but their cumbersome spidery shape makes transport over long distances uneconomic. On site chipping is impracticable because of the inclusion of flints and other foreign matter. The cleaning of stumps prior to chipping is essential. Crushing with giant rollers has been attempted in some eastern European countries, but it has not yet been tried in Britain. In a few countries including Finland and Poland stumps and roots are included in the raw material for pulp and board manufacture. In Poland the use of pure stumps from sandy soils increases wear and tear on hardboard production lines because of fine silica particles that adhere to the chips made from roots, even after they have been washed.

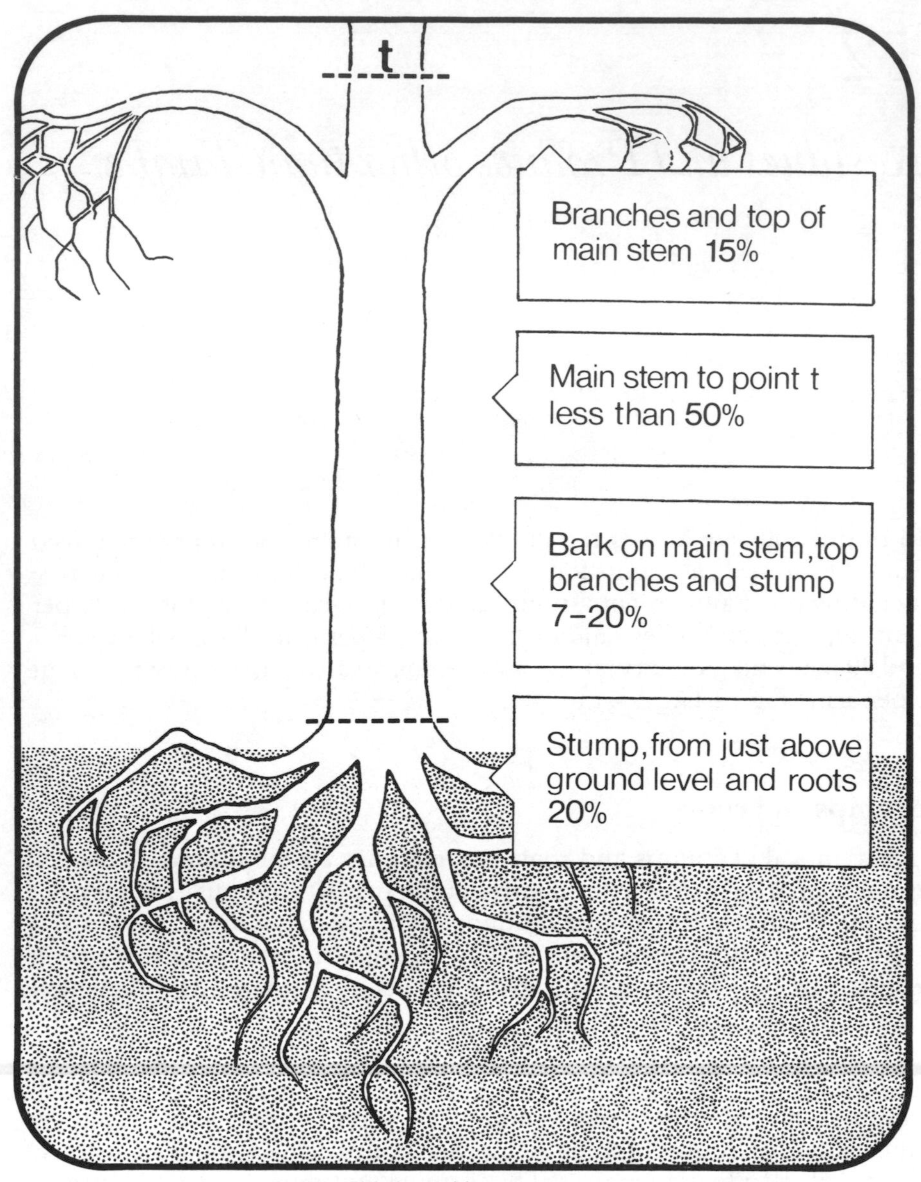

Fig 12.1 Whole tree bio-mass.

Branches and tops

The main use for branches is a small one. The foliage of some species is in demand for floristry (i.e. the enhancement of arrangements of cut flowers) (Fig 12.2) and for wreath-making; for both of these purposes sales are normally by weight. It is not possible to generalise on the requirements of this market except to note that the preferred species are those conifers which retain their needles well, such as the silver firs, Lawson cypress, western red cedar and western hemlock, but Douglas fir and European (Norway) spruce are also used, especially for wreaths (Fig 12.2). Broadleaf species such as holly and beech may be in demand in some localities, but care is required to prevent unsightly wilting in the latter. Unusual effects can be obtained with Eucalyptus foliage.

During November and December the tops of most of the evergreen coniferous trees find a ready market as Christmas trees, even though they may be inferior in shape to purpose grown crops. Tradition requires that the bulk of this market be met by European spruce, but a better performance in respect of needle retention is given by the silver firs. Scots pine is sought in the New Forest area, and despite the sharpness of its needles, Sitka spruce is used in some regions. It makes commercial sense to time a part of the year's thinning programme of the acceptable species to coincide with the period of optimum demand.

Fig 12.2 Holly, European spruce, Lawson's cypress used in floristry

The harvesting of whole trees, which facilitates the chipping of tops and branchwood, is sometimes undertaken using a tractor-driven portable chipper. Chipping the tops and branchwood may, in fact, be carried out to reduce the fire hazard at roadside or forest depot, even though a market might not exist for the chips (Fig 12.3). After conventional harvesting the residual lop and top may be a hindrance to mechanised site preparation such as screefing and drainage; if so chipping and removal of the lop and top may be undertaken. However, there may be reasons for retaining the branches and tops within the stand of trees; these include the recycling of nutrients contained in the foliage, the reduction of erosion and dessication of the soil and the suppression of some weed growth.

The usefulness of the chipped tops and branchwood depends upon the proportion of wood to bark and on the species; if the wood content is relatively low then the uses are restricted to those of land-fill, cattle litter, coarse landscaping and mulching. The moisture content of such chips is usually too high for them to be used as fuel except in specially designed furnaces, but even so it is usual to allow some drying to take place before chipping.

Fig 12.3 Tractor mounted wood chipper.

The chips can be processed into board or into pulp, but for both of these outlets it may be necessary to enrich the yield by feeding some bark free wood into the chipper. However, one hardwood pulp mill purchases chips, which include bark, prepared at sites distant from the mill. The acceptability of chips produced in the forest for particle-board or other building board varies considerably, not only with the individual company, but also with the price and availability of technically superior alternatives such as dry joinery and sawmill waste. Forest produced chips are quite unsuitable for the manufacture of high quality carton of the type used for the packaging of frozen food. Forest chipped wood is used for the surfacing of "all-weather" racecourses, but has to compete with dry wood waste from joineries and sawmills.

Wood chips that have been stored for some time can give rise to the disease known popularly as "farmer's lung" (see also page 174) and for this, and to improve their rate of drying, the tendency is to find an alternative to chipping which will produce larger pieces than the conventional knife chipper. The development of machinery to make so-called "chunks" is under way in Sweden and elsewhere.

Bark

Every cubic metre of stem wood harvested carries with it a fifth to a fifteenth of a cubic metre of bark, and bark has become the most important by-product of the forest, being eagerly sought for horticulture, landscaping the surfacing of play areas and equestrian purposes. Prior to 1970 no major uses for conifer bark had been developed in the UK and the once important oak tan-bark industry had dwindled to a few hundred tonnes of oak bark for the manufacture of top quality leather. Earlier attempts to find an economic outlet for bark had not been successful. In the 1950s a match factory attempted to supply shredded poplar bark for mulching, but had found it to be unprofitable; and during the late 1960s a wood pulp mill undertook the compressing of mixed conifer bark into fuel briquettes for household use, but found the intermittant seasonal demand to be a major problem. Other pulpmills burned bark to raise steam but the processes then used were not efficient in energy terms.

During the 1950s the Forestry Commission and the British Leather Manufacturers' Research Association undertook extensive development work to assess the prospects for the use of conifer bark by the home leather industry since bark contains a number of potentially useful extractives, principally tannin. The initial results were promising. It was confirmed that the spruces and larches had a high percentage of tannin (sometimes exceeding 20 per cent of the dry weight when extracted with methanol), and that they could be used for the production of serviceable grades of leather. The actual tannin contents of the bark of the different species is given in the table overleaf.

Other work on bark included a detailed examination of Douglas fir bark which is of particular interest because in addition to tannin, it contains two other extractives of commercial value, namely wax of a type suitable for the manufacture of polishes (when extracted by an organic solvent the yield is about 6.2 per cent) and dihydroquercetin which has pharmaceutical applications (hot water extraction results in a yield of about 2.0 per cent of the dry weight).

This research work on extractives failed to stimulate any commercial

Species	Tannin % of dry weight	Other Water solubles % of dry weight	Red Colour Intensity*
Oak bark	13.0	—	6.9
Sweet chestnut bark	15.2	11.9	6.0
Hazel bark	6.3	6.4	9.0
Birch bark	6.7	9.9	2.4
Scots pine bark	11.7	13.8	1.6
Corsican pine bark	12.4	15.9	1.3
European larch bark	12.5	9.0	5.3
Japanese larch bark	16.5	9.4	4.8
European spruce bark	13.7	14.3	3.6
Sitka spruce bark	17.5	12.2	4.9
Douglas fir bark	9.9	—	—
Western hemlock bark	16.7	11.0	6.4

* Units registered on a standard colorimeter.

developments for a number of reasons. These included the low costs of synthetic tannins and technically superior vegetable tannins from tropical regions and the high cost of harvesting bark by hand. It was discovered that the machine barking undertaken at pulpmills in the UK and in the preparation of pitprops not only reduces the tannin content, but it also raises the colour of the extract to levels which would produce unacceptably dark leather. The bark must therefore be stripped by hand and air-dried before delivery to prevent deterioration resulting from mildew.

It is in horticulture and the associated landscaping where bark has found its widest application.

The role of bark in nature is to protect the tree from dessication and from attack by harmful fungi and insects as well as from larger animals. To fulfill these functions it has evolved into a durable water-resistant tissue containing substances which are repellant or unpalatable to pests. These are properties which make it suitable for many horticultural uses provided that it is properly prepared. Fresh bark cannot be used beneficially in close contact with roots; this is because it contains a small quantity of volatile oils called monoterpenes which inhibit plant growth. The preparation of bark for horticultural purposes, therefore, involves a fermentation process which reduces the monoterpenes to innocuous levels or even to lower concentrations at which they actually stimulate growth. The bark from the barking machine is first pulverised and then stacked in heaps of several tonnes for a minimum of six weeks. During this period it is rapidly colonised by thermophillic bacteria which raise the temperature to about 80°C.

This fermentation process is referred to as "maturing" or "weathering" or, incorrectly, as "composting" and has three beneficial effects:

- the monoterpene content is reduced by 75 per cent, hence plant growth is no longer affected adversely.
- the sustained high temperature ensures that all parasites and weed seed are destroyed.
- the somewhat acid pH at 4.5 is raised to the horticulturally desirable level of about pH 5.6.*

While the development of selective herbicides has brought about a lessening of interest in mulches as a way of supressing weeds, mulches are, nevertheless, beneficial in that they help to conserve soil moisture by reducing surface evaporation without impeding unduly the downward percolation of rain, as well as insulating the roots from abrupt changes in temperatures. Furthermore they often create a pleasing contrast with foliage, flowers and lawns. This aesthetic effect can be varied by changing the species of bark or the particle size (for example a "peat" effect can be obtained by using finely pulverised spruce bark; alternatively an appealing rugged appearance can be achieved by mulching with chunky pine bark).

While it is possible to strike cuttings in bark which has been pulverised and fermented and also to pot and grow a wide range of plants in it, the best results are obtained by blending bark and peat together with a little sand or grit. However, bark cannot contribute significantly to the nutrition of the crop and the addition of a compound high-nitrogen fertiliser is essential if nutrient deficiencies are to be avoided. Although it should always be borne in mind that the main role of bark in potting mixtures is to provide good aeration to the root system, it is now known that root hairs are capable of penetrating within the bark particles thus improving the plant's anchorage.

Other horticultural uses include orchid culture for which a coarse grade of pine bark is usually preferred; mushroom culture where a finely pulverised spruce or mixed conifer bark has proved most suitable; and bulb-forcing for which the addition of lime is required.

Following an increased awareness of the risk of injury to children there has been a growing demand for bark for spreading over the surfaces of play areas. For this use larger particles of bark, roughly between 50 and 70 mm in size are spread beneath swings, slides, climbing frames and other

* A pH of about 5.6 is ideal for most horticultural crops, at this level the vital trace elements are readily available to the plant. At lower levels the availability of molybdenum diminishes and manganese toxicity may occur. At higher levels the availability of iron, manganese and magnesium is reduced and deficiencies may occur. Nevertheless some crops thrive at a higher pH around 7.0.

Fig 12.4 Bark in play area provides soft landings.

items of play equipment to ensure that a falling child experiences a "soft landing" (Fig 12.4). A relatively small volume of bark is used for equestrian purposes such as the surfacing of gallops, manèges and show jumping arenas; for these purposes it should have an average particle size of about 50 mm. It serves particularly well in totally enclosed areas because it has the useful property of absorbing ammonia and other foul odours. If used as a deep litter for poultry good ventilation is essential, otherwise the chicks can contract a serious illness called asperigllosis as a result of inhaling the spores of the fungi growing on the bark (there is a similarity between this disease and "Farmer's Lung" caused by fungi growing on stored wood chips, see page 171).

Bark can be used as fuel and the most modern techniques make it economic to use it in pulp mills for steam-raising (this is standard practice in Scandinavia). Nevertheless there are some disadvantages. The moisture content must be reduced to 60 per cent of the dry weight (in freshly felled coniferous bark it is usually greater than 110 per cent). This is often achieved by squeezing out the water mechanically rather than by applying heat. Bark has a much higher ash content than conventional fuels (at five per cent it is at least twice that of most species of wood) and the ash has a low slagging temperature, which means that specially designed furnaces may be required unless the bark is mixed with other fuels with lower ash contents.

Developments overseas in which bark is used together with other residues in the production of building boards, and where the tannin in the bark is condensed into an adhesive to provide the necessary cohesion, have not been pursued in Britain. However, during the 1970s a company in Somerset was engaged in the production of cement building blocks containing a high proportion of bark. This commodity had good heat and sound insulation properties.

Bark produced by the radially mounted blade type of barking machine (commonly used in the preparation of pit props and fence posts) invariably contains a large proportion of wood (approximately 40 per cent) and this has limited its usefulness.

It had been tried for mulching and landscaping, but it has an untidy appearance compared with wood-free bark. It has also served as litter for cattle. It is sometimes purchased for the core layer of particle board. For this use it is fragmented and enriched by the addition of other wood residues such as planer shavings.

Wood residues

A brisk trade exists in the marketing of residues from sawmills and wood-working plants of all kinds, with a well organised body of wood-waste merchants involved in their acquisition, transport and sale. The residues involved include sawdust, off-cuts, slabwood (with or without the bark) and planer shavings together with other joinery waste. All of these by-products can be used by the wood chipboard and boardmill industries, but are not suitable for orientated strandboard or high-quality cardboard production. The acceptability and the price obtained varies with the moisture content, the species (or mix of species) and the degree of contamination with non-wood substances. Some chipped residues are exported to pulp mills in Scandinavia.

Shavings and sawdust are in demand as deep litter for poultry, but dryness is of paramount importance. As already noted (above) moist litter facilitates the development of fungus on the surface which causes an ailment in poultry called aspergillosis. It is also important that residues from wood which has been treated with a preservative should not be used for deep litter as this can cause the tainting of poultry flesh.

A small quantity of broadleaf residues is sometimes purchased by the hardwood pulp and the charcoal industries and some charcoal manufacturers also accept softwood off-cuts.

13

The Chemistry and Structure of Wood and their Implications for Pulp Production

A knowledge of the constituents of wood cells together with an appreciation of the nature and arrangement of their molecular components adds to an understanding of the performance of timber in service, and may well give an indication why a species may be preferred for a particular end use. Nevertheless the information in this chapter is necessarily restricted to a mere outline of the anatomy and molecular structure. Students and others requiring an exhaustive account are referred to the existing excellent publications such as *The Anatomy of Wood* (Wilson and White), *The Structure of Wood* (Jane), and for hardwoods *The Anatomy of the Dicotyledons* (Metcalfe and Chalk).

Fundamentally, wood consists of compounds of the elements carbon, hydrogen and oxygen, primarily as components of the molecules of cellulose,* lignin and hemicellulose (the cellulose and hemicellulose are collectively termed holocellulose).

Cellulose

Cellulose accounts for roughly half of the dry weight of wood. It is formed as a result of glucose molecules being condensed (polymerised) into long unbranched chains consisting of several thousand glucose "units". The width of the individual cellulose molecule is far below the limit observable under the microscope; hence, details of its structure and organisation within the cell wall have been obtained by other techniques, especially the

* Some authorities refer to a-, b-, or y-cellulose: these refer to the fractions of holocellulose classified according to their solubilities in alkali. This appellation is best avoided in the light of modern knowledge, β- and α-cellulose are in fact components of hemicellulose. [α-cellulose is insoluble in a 17.5% aqueous solution of sodium hydroxide, β-cellulose is soluble but reprecipitated on acidification, while α-cellulose is soluble but not reprecipitated on acidification].

use of polarised light, electron microscopy, and X-ray diffraction. While cellulose is the groundstock for the manufacture of paper, its other industrial uses are too numerous to list effectively, but man-made textiles and explosives should be given special mention. It is also worth noting that during the Second World War a number of plants were built on the European mainland for the acid hydrolysis of cellulose into fermentable sugars for the production of alcohol for motor spirit, with yeast for animal fodder as a by-product. (In some eastern European countries hydrolysis of the waste liquor from pulping is undertaken for the culture of yeast; the sugars involved are pentoses which do not yield alcohol on fermentation).

Hemicellulose

Hemicelluloses are a group of compounds with large molecules containing up to 200 sugar "units", formed by the union of glucose and other sugars with uronic acids* to produce branched molecular chains much smaller than those of cellulose. They are somewhat less stable chemically, being readily degraded by both acids and alkalis; for this reason species with a high proportion of hemicellulose have a low acid resistance. The implications for home-grown timbers is that softwoods with a relatively small hemicellulose content are better suited for use in contact with acids than hardwoods. Because hemicellulose usually contains a number of pentose "units" (i.e. sugars with five carbon atoms — one less than glucose), which can be processed with acids to produce the important industrial solvent furfural (furfuraldehyde), it is sometimes suggested that hardwood residues might be used as a raw material for its production, but this has not been attempted commercially because alternative plant residues are available at a lower cost.

Lignin

Lignins are deposited within and between the cell walls giving the wood tissue its rigidity and contributing greatly to the bonding. They are bulky non-fibrous molecules comprising largely a network of "units of six-member carbon rings each with a three-member side chain". Although lignins are fairly acid resistant, they are rendered soluble by alkalis, sulphites and sulphates — a property used in chemical pulping processes whereby much of the lignin is washed away to leave a high-cellulose pulp. It softens at elevated temperatures; a property which facilitates the steam bending of wood, and also the separation of tracheids in mechanical pulping.

* Uronic acids are formed by the oxidation of the primary alcohol groups in sugar molecules, and are named after the sugars from which they are formed.

The lignaceous liquor produced as a by-product in chemical pulping has a limited number of uses depending upon the pulping process employed. In the alkali sulphate or Kraft process and in the sulphite processes (none of which are undertaken in the UK) it is fed into the furnace to provide energy and to facilitate the recovery of chemicals for re-use. Additionally, ligno-sulphite liquors have been used for tanning and for core-binding in foundry work.

Other substances

As already noted in Chapter 4, approximately half the weight of the wood in freshly felled trees is accounted for by water. In the sapwood, where the tissue is still living, not only are the cell walls saturated but water is an integral part of the protoplasm. In the undried heartwood there is rather less water in the cavities of the now dead cells, but moisture is retained in the walls duly occupying the voids between the structural elements, complementing their mechanical function of stem support.*

The main coniferous species contain resin, which although it has never been recovered commercially in Britain, is the source of a number of useful products collectively called "naval stores". Naval stores are obtained by the distillation of resin to give the useful products of rosin (the residue) and oil of turpentine as the distillate. The turpentine may then be refined further to yield compounds such as α-pinene and β-pinene which have a number of pharmaceutical and industrial applications. These products are monoterpenes similar to those found in the bark of the same species (see chapter 12). The constituents of the turpentine and their relative proportions are characteristic of a species or even of a provenance within a species, and are used by research workers for "genetic finger-printing".

Tannins occur in significant quantities in some hardwoods; the heartwood of sweet chestnut contains about five per cent of the oven dry weight and that of oak about one per cent; in eastern Europe tannins are extracted from the chipped wood at pulpmills for use by the leather industry. Tannins contribute substantially to the good natural durabilities of oak and sweet chestnut, but they have the undesirable effect of both corroding ungalvanised steel nails and screws, and causing unsightly black stains to develop around them, especially in damp situations.

The structure of coniferous wood

The dominant component of softwood tissue is the tracheid, which has the functions of conducting water and dissolved salts from the roots to the crown, and of affording mechanical support to the foliage. The tissue

* The water retained in the cell wall is known to play an important role in binding the constituent parts together, especially by forming non-covalent hydrogen bonds between the cellulose microfibrils and the matrix (of the lignin and hemicellulose) as well as within the matrix. Water can therefore be regarded as an essential structural component of the cell walls in the wood tissue.

formed during the early part of the growing season is termed the *earlywood* (or spring wood); it is primarily concerned with water conduction, and it consists of relatively large diameter thin-walled tracheids. Tracheids formed towards the end of the growing season are narrower with thicker walls to provide stronger support for the stem during the forthcoming winter; this tissue is called *latewood* (or summer wood). In some species such as Scots pine and the spruces the transition from earlywood to latewood is gradual, while in others such as Douglas fir and the larches it is abrupt, making the annual rings very conspicuous on end-grain surfaces and giving a striped appearance to longitudinal surfaces. The tracheids (referred to as fibres by the paper technologist) are tube-like structures, but their ends are shaped like a rounded chisel appearing pointed in tangential sections and rounded in radial sections. In the narrower elements of the latewood the rounded shape is less obvious. They are cemented together by a lignified hemicellulosic layer called the *middle lamella*. The tracheid wall usually consists of four distinct layers, rather like sleeves within sleeves, but they can be distinguished from each other by the use of the polarising microscope or the electron microscope, the salient difference between them being the orientation of the bundles of cellulose chains, called *microfibrils*,* which become manifest as different intensities of illumination when viewed under polarised light. The actual orientations (Fig 13.1) are as follows:

- The outermost layer called the *primary wall*, which adjoins the middle lamella, has its microfibrils loosely arranged almost at random within a hemicellulose network. At maturity it has a high lignin content. It is the only wall of the tracheid during its period of growth to its full size.
- The second layer is known as the *S1 layer of the secondary wall*. Its microfibrils are arranged in shallow spirals in both the left hand and right hand directions, lying at angle of some 15–40° to the transverse.
- The third layer is called the *S2 layer of the secondary wall* and has its microfibrils arranged spirally at a steep angle to the transverse, usually about 60–80°. The microfibrils of this layer are more closely packed than elsewhere, and since the S2 is thicker than the S1 and S3 layers, the average orientation of the microfibrils in the wall as a whole is very near that of the S2 layer alone, ie 60–80° to the transverse. This has a bearing on the shrinkage and strength properties of the wood.
- The innermost layer is called the *S3 layer of the secondary wall*, it has microfibrils spiralling at a shallow angle between 10–30° to the

* The arrangement of the cellulose chains within the microfibrils results in a structure which is largely crystalline and it is this which facilitates investigation by polarised light and by X-ray diffraction.

transverse. It is not formed in the spruces.

– The cell cavity is called the *lumen*.

The deposition of lignin within the holocellulose framework of the wall decreases with the distance from the middle lamella, being most intense in the primary layer and the S1 layer. Nevertheless, because the S2 layer is usually much thicker than the others it contains most of the lignin.

The other wood tissues are not lignified in the sapwood but usually become so in the heartwood; they comprise soft tissue or *parenchyma* in the rays and the pith, and in some species, thin walled cells lining the resin canals called *epithelial cells*.

The passage of fluids between wood cells is principally through minute pits. Pits connecting parenchyma cells are normally simple gaps in the cell wall, but with the intercellular middle lamella remaining intact (see Fig 13.2), but those between tracheids are much more elaborate and are called *bordered pits*; they are found almost exclusively on the radial walls. Bordered pits display a variety of forms and arrangements — often characteristic of a genus, and they are thus a most useful feature in the microscopic identification of softwoods. A further interesting facet is that they serve as valves; they have a thickened central disc* which is deflected tangentially when the cell dries out so as to plug the pit and impede the further passage of moisture; in this condition the pits are described as

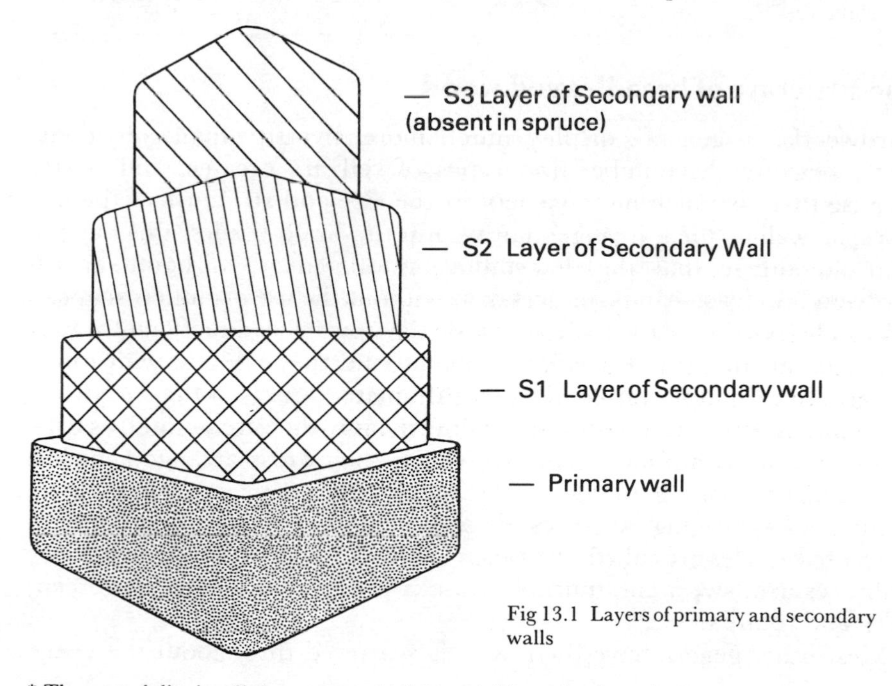

Fig 13.1 Layers of primary and secondary walls

* The central disc is called a torus, and it is not visible under the optical microscope in western red cedar and in some Southern Hemisphere conifers.

aspirated (see Fig 13.3). In some species such as spruce or Douglas fir, the aspiration of the pits explains why these timbers are easily treated by a diffusion process in the green state, but are difficult to impregnate with preservatives under pressure after they have been dried.

The pits connecting the parenchyma with the tracheids (called *cross-field pits*) are also useful in the identification of softwoods, and in some instances they facilitate the separation of groups of species within a genus when radial sections are viewed under a microscope.

The positioning of the vertical (axial) resin canals are also an aid to identification, which can be used to a limited extent when no microscope is available. For example, in the home grown pines these canals are relatively large, and their distribution, concentrated in the latewood, can be observed with a hand lens; this compares with much smaller (barely visible with a hand lens) canals in the larches, which appear to have a random distribution.*

The coniferous genera can also to some extent be identified by noting the size, shape and distribution of the rays when observed in tangential sections, making particular note of the presence or absence of radial resin canals.

Additionally, horizontal ray tracheids, with characteristic bordered pits, may be present in some genera, but these are best viewed in radial sections.

The structure of broadleaved wood

Hardwoods, in general, display much more diversity than softwoods, partly because the number and types of cell are greater, and partly because there is much more variety in the size and structure of the cell types, as well as their arrangement within the wood tissues; this has the clear advantage that the determination of genera, of home grown hardwoods at least, can be undertaken without the use of a microscope.

Vessels (pores) are the main units for the upward conduction of water, while they are the largest cells in the wood tissue, they exhibit a wide range of sizes. In ash, for example, the large diameter vessels of the earlywood are readily seen by the unaided eye, while in holly they are so small as to be barely visible with a hand lens. However, their gross arrangement is a useful starting point in the identification of genera or even species.

Timbers exhibiting large vessels appearing on end grain surfaces in concentric circles are called *ring-porous* (Figs 13.2 and 13.3). This category embraces ash, sweet chestnut, elms, oaks [other than holm (evergreen) oak], and robinia.

Most other genera have their vessels scattered throughout the year's

* Resin canals are present in all major British softwoods, but are normally absent from western hemlock and the silver firs.

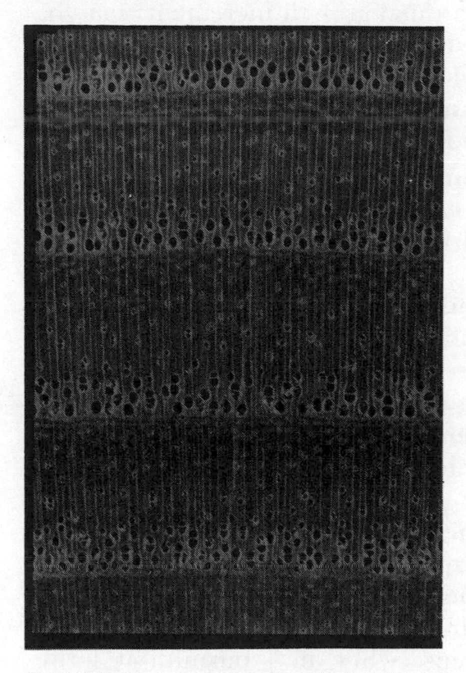

Fig 13.2A Ring porous hardwood ash × 5
End-grain view of ash (*Fraxinus excelsior*) a ring porous hardwood where the annual growth rings appear as a continuous ring of larger vessels which are readily seen with the unaided eye. The fairly uniform rays are narrow and difficult to observe without a hand lens.

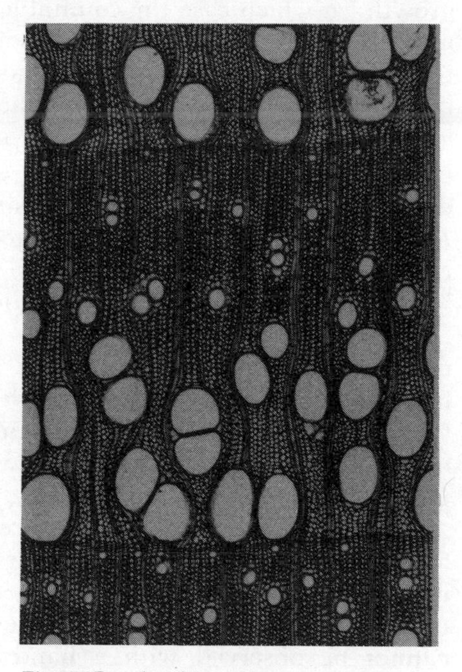

Fig 13.2B Ash × 24
Transverse section of ash (*Fraxinus excelsior*) showing contrast between earlywood with large vessels and thin walled fibres, and latewood with sparse smaller vessels amid thicker walled fibres.

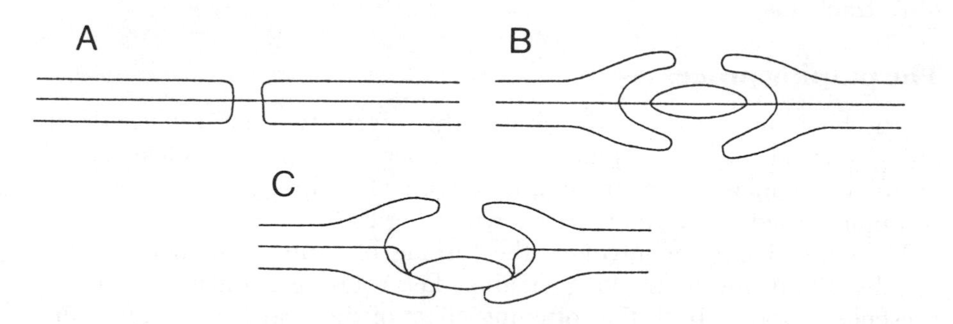

Fig 13.3 Diagramatic representation of a section through a simple pit. (A) Diagramatic representation of sections through bordered pits in (B) the "normal" and (C) aspirated states. [note the blocking of the aperture by the torus in the latter].

growth (in which case the countable annual growth increments may be marked by bands of parenchyma [soft tissue] as in sycamore). These are termed *diffuse porous* (Figs 13.4 and 13.5). They include alder, apple, beech, birch, cherry, holly, hornbeam, lime, London plane, maples, *nothofagus*, pear, poplars, sycamore, and willows.

Walnut, which has relatively small vessels arranged in vague discontinuous rings, is regarded as *semi-ring-porous*; and most species of *Eucalyptus* have their vessels arranged in short rows oblique to both the radial and tangential planes.

In some species notably the pedunculate and sessile oaks, the vessels of the heartwood are obstructed by bladder-like bodies called tyloses, making them quite impermeable to fluids. This explains why the heartwood of these species was in demand for tight cooperage, and why the sapwood is unsuitable* (the American red oaks principally *Quercus rubra*, and Turkey oak, *Q. cerris*, which do not form tyloses are also unsuitable).

The distribution of soft tissue (parenchyma) as seen in cross-section can also be a key factor in identification, especially with regard to its position in relation to the vessels. In some species such as alder it is scattered fairly uniformly through the wood tissue independent of the vessels, and it cannot be observed with a hand lens; while in laburnum it forms conspicuous bands connecting the vessels tangentially.

Fibres are concerned with the function of support, and they generally form the bulk of the wood tissue; but they have few distinguishing features and are not used in macroscopic identification. The same is true of fibre tracheids which are found in some hardwoods. Hardwood fibres have an ultrastructure similar to that of conifer tracheids, except that bordered pits are absent (if bordered pits were present they would be redefined as fibre-tracheids).

The pulping processes

At the beginning of the 20th century there were two main methods of pulping wood for the manufacture of paper and other cellulose-based products, namely mechanical and chemical pulping. The processes commonly used in the 1950s are shown in Fig 13.6.

Mechanical pulping involves the forcing of round softwood billets, usually about one metre long, against fast-rotating grindstones in the presence of water. Both the softening effect of the heat from the friction and the abrasive action of the grindstones reduces the wood to a coarse

* Other than for the production of whisky, wooden barrels have virtually disappeared from the home drinks industry. Very little British oak is used for cooperage, because most of the maturing takes place in old sherry casks or in barrels remade from second-hand bourbon whisky casks. Imported American white oak, mainly *Quercus alba* is generally used for their repair.

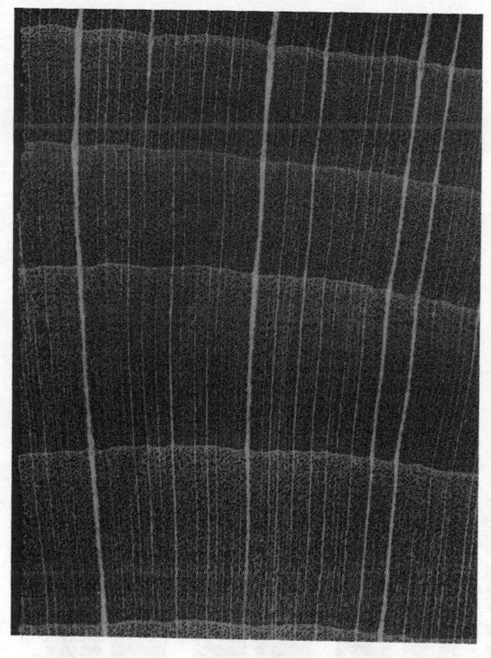

Fig 13.4 Diffuse porous hardwood Beech × 9
End-grain view of beech (*Fagus sylvatica*) a diffuse porous hardwood where the annual growth rings are manifest as terminal bands of fibres. The small vessels are not visible to the unaided eye. Note large rays of varying widths.

Fig 13.5 Beech × 24
Transverse section of beech (*Fagus sylvatica*) showing vessels of fairly uniform diameter randomly arranged in a matrix of fibres. Note the dense thicker walled fibres in the latewood.

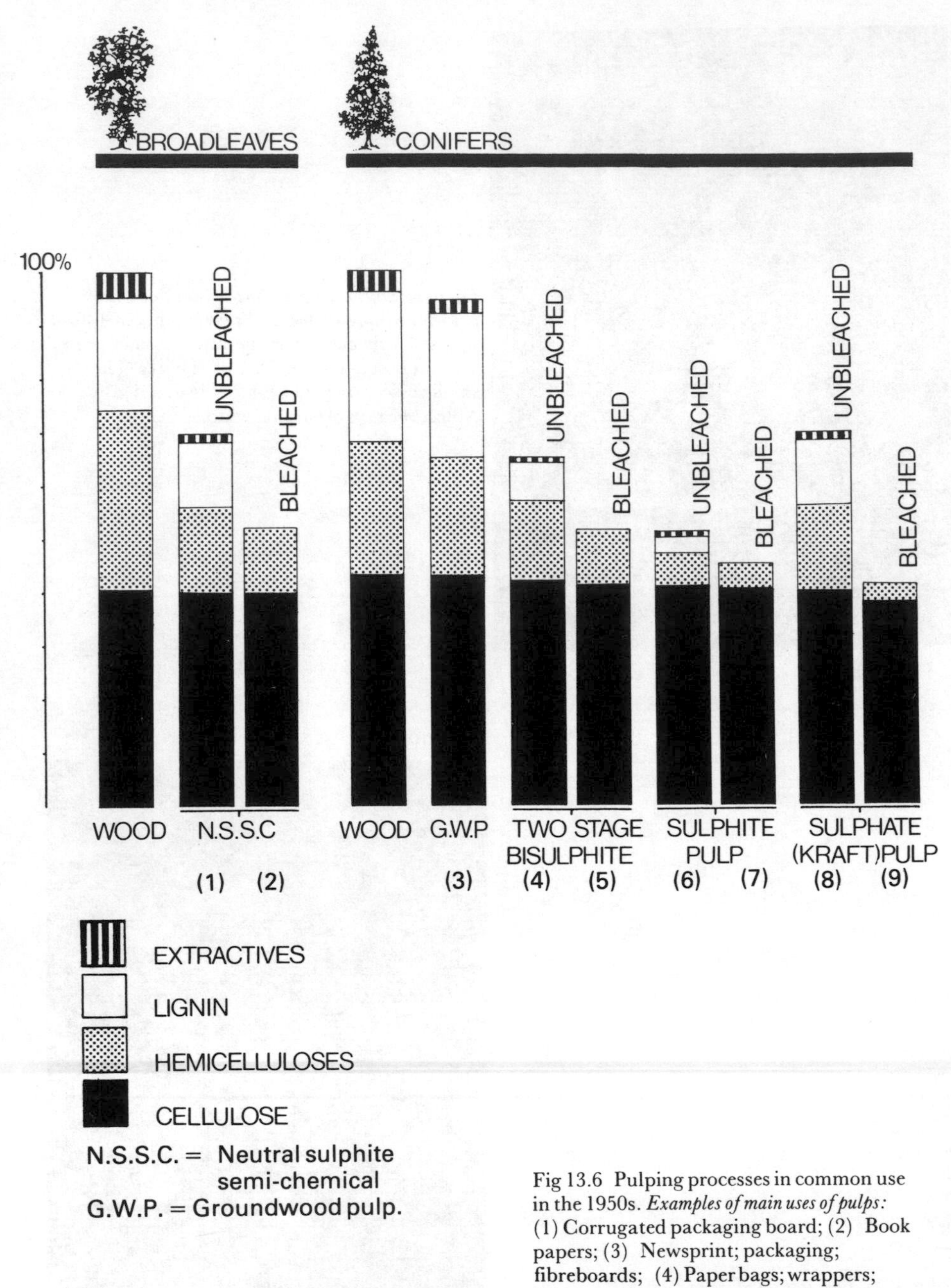

Fig 13.6 Pulping processes in common use in the 1950s. *Examples of main uses of pulps:* (1) Corrugated packaging board; (2) Book papers; (3) Newsprint; packaging; fibreboards; (4) Paper bags; wrappers; (5) Newsprint; writing and wrapping papers; (6) paper bags; (7) Writing and wrapping papers; (8) Strong brown and other wrapping papers; (9) Writing and wrapping papers

pulp consisting of tracheids, minute bundles of tracheids and torn fragments of tracheids. Spruce is the preferred species for grinding, but other conifers (except larch which is too resinous and Douglas fir which is too dark) have been used in mixture with spruce. This process has the advantage of high yields — usually exceeding 90 per cent, and a relatively low requirement for water and chemicals resulting in only a small effluent problem. However, the pulp produced, which is used mainly for the manufacture of newsprint and other printing papers, has low strength properties (due to the tearing of tracheids during grinding)* and is lacking in brightness; it is therefore normal practice to add up to 20 per cent of a bleached chemical pulp to provide adequate strength. A white filler such as china clay may also be added to some grades of paper to improve the colour and to provide a smooth surface for printing. Specially thin printing papers are now being produced with a fine plastic-coated surface to give improved printing properties.

The pulp mills in Britain which used the stone groundwood process to make mechanical pulp were closed in the early 1980s. Groundwood production by the disc-refiner process — where pulping is effected by forcing wood chips through a minute gap between two abrasive steel discs one of which is rotating rapidly — continues and has been expanded. If heat is applied as super-heated steam the process is called thermo-mechanical pulping (TMP). In this method the elevated temperature has the effect of softening the lignin in the middle lamella to facilitate better separation of the tracheids to yield a higher grade of pulp.

The product is more uniform than that produced by stone-grinding. Among its main uses are the production of carton (known in the industry as board) for such purposes as the packaging of frozen food and cigarettes. The carton is usually lined on one or both sides with a bleached chemical pulp which improves both its appearance and its strength. A further advantage of the disc-refiner process is that because the raw material takes the form of chipped wood, a wider range of lengths and diameters of softwoods can be used. Here too, spruce is the preferred species but some other conifers including *small proportions* of larch and Douglas fir may be included.

Groundwood pulp is now also being produced using the pressurised groundwood process (PWG). This process uses the traditional groundwood method of grinding logs against millstones in a pressurised atmosphere. This softens the wood without breaking the tracheids. Essentially the pressurisation suppresses boiling and thus precludes the disruption of the water film between the wood and the grindstones. This film acts as a lubricant. The process results in a stronger pulp which has

* The strength of a pulp depends largely on the ratio between the length of the tracheids or fibres and their diameters. The strength of the paper produced is assessed in a number of ways including resistance to tearing and resistance to bursting in both the wet and dry condition.

several advantages from the papermakers point of view, and which is fully competitive with thermo-mechanical pulp.

Chemical pulps are made by treating (digesting) wood with compounds such as caustic soda, sodium sulphate or calcium sulphite, usually under pressure and at an elevated temperature, so that the lignin is made soluble and can be washed away. As a general rule the more drastic the digestion process the lower the yield, but the higher the quality (in terms of strength and uniformity) of the pulp and paper produced.

Where the appearance of the product is of little importance, the pulp may be used without bleaching for the manufacture of commodities such as brown paper, but where the end product is to be used for writing, printing or the wrapping of food, the pulp is bleached. Bleaching has the effect of depressing the yield, and once again the more drastic the process the lower the yield. In fact to improve the whiteness while preventing an unacceptable loss of yield, bleaching may be carried out in up to five stages, the pulp being thoroughly rinsed between each stage. Chemical pulps are of uniform good quality with superior strength properties, but their manufacture has a high water requirement — in some cases 90 tonnes of water per tonne of dry pulp. Some processes can produce environmentally undesirable effluents and others may have an odour problem.

The softwood chemical pulp mills which were operating in Britain in the 1960s and 70s have now closed; their small capacity made them uncompetititve against the very much larger mills in North America and Scandinavia, especially during periods when a strong £ Sterling made imported pulp relatively cheap.

Developments in pulping technology during the past four decades have included efforts to combine the high yield of mechanical pulps with the superior strength of the chemical pulps. In these methods known as semi-chemical pulping, chipped wood is first softened by partial delignification using a chemical such as sodium bisulphite, after which it is ground to pulp by a disc-refiner. Two mills of this type are operating in Britain using hardwoods.

The product is an unbleached pulp* used for the manufacture of packing materials such as corrugated board. However, if semi-chemical hardwood pulps are bleached, they can be blended with conifer pulps to produce writing, wrapping and printing papers, where they have the effect of improving both the opacity and the printing properties.

Technically pulp suitable for a specific grade of paper can be made from wood taken from any part of the tree, including the roots. However, the

* The fibres of hardwoods are shorter than the tracheids of softwoods, hence the pulp which they produce has different strength properties from conifer pulps; in particular, hardwood pulps give good compressive strength when used to make the corrugated element of the familiar corrugated packing carton.

specification for an individual pulpmill will be determined by a combination of factors, whose priority will depend on the type of pulp to be produced. In an integrated mill where paper (rather than pulp for sale on the open market) is the end product, the availability of pulp from other sources or waste paper for blending with the mill's own pulp will also have an influence.

The factors referred to above include:

- the intrinsic qualities of the wood; for example for strong wrapping papers and newsprint where resistance to tearing is important, long fibred softwoods are required; for corrugated packing media and some fine writing papers hardwoods are in demand. The presence of extractives (e.g. tannin in oak and sweet chestnut and resins in the larches) also can affect the performance of pulp.
- the need in most pulping processes to remove the bark. Debarkers require wood with a certain degree of straightness and a minimum diameter.
- the minimum and maximum diameters and lengths that can be accommodated by the chippers, or in the manufacture of stone groundwood pulp the dimensions of roundwood which can be presented to the grindstones.
- the ranges of lengths and diameters that can be handled by the conveyors (e.g. to and from the debarker, and to the chipper).

Whatever the ideal specification for a particular pulp mill may be, it must be tempered with the reality of the potential supply available within an economic catchment area and by the relative costs of harvesting and transporting the wood from stump to mill gate.

Pulp mills also have to take into account the competition from the sawmilling industry which can make more profitable use of roundwood above a certain diameter and length. The critical dimensions do, of course, depend on the relative prices for pulp and for sawn timber and the economic dividing line between pulpwood and sawlogs varies over time.

For softwoods, lengths in the range 1.5 to 3.0 metres, with diameters ranging from 6 to 40 cm, are common in Britain. Spruce is the preferred species for mechanical pulp, but some other conifers are accepted in limited quantities. (Western red cedar is usually avoided, not only has it a poor colour but it can also corrode equipment).

In hardwood pulping in Britain the end product can tolerate virtually any species of broadleaf; and as the wood is used unbarked, it can be accepted in a wide range of sizes.

Most pulping processes prefer fresh felled wood and purchase by weight so as to encourage prompt delivery after felling. This also promotes forest hygiene by encouraging rapid removal of felled material (Chapter 8).

Pulping processes employed in Britain and the species of pulpwood used

The following four examples demonstrate the range of pulping processes used in Britain and the species and sizes of wood used:

St Regis Paper Co Ltd, Sudbrook Mill, Gwent

St Regis Paper Company produces 110,000 tonnes a year of semi-chemical fluting medium which is made from 65–70 per cent neutral sulphite semi-chemical pulp. The balance of the furnish is waste paper.

The pulpwood specification is for broadleaved species, in 2.3 to 2.4 metre lengths, with a diameter range of 2 inches (50 mm) (minimum) and 18 inches (440 mm) (maximum). A log splitter was installed recently which will enable the mill to abandon the upper diameter limit.

Caledonian paper plc, Irvine, Ayrshire

Caledonian Paper makes a light-weight coated paper for use in colour magazines and brochures. The mill capacity is 170,000 tonnes of paper per annum.

About half the pulp requirement is made at the mill from British grown spruce pulpwood using the pressurised groundwood process; the other half is imported chemical pulp. The pulpwood specification and details of the wood handling process are shown in Figs 13.7 and 13.8.

Shotton Paper, Clwyd

Shotton Paper makes newsprint using the thermo-mechanical process. The wood used is 90 per cent spruce; the remaining ten per cent may be pine, grand fir or western hemlock. Sawmill residues may be used up to 15 per cent of the total wood consumption. Waste paper (e.g. over-issue newspapers, periodicals and magazines) is also used to supplement the fibre from pulpwood.

The mill capacity is 400,000 tonnes of newsprint a year.

Thames Board Mill Ltd, Workington, Cumbria

Thames Board uses the disc-refiner groundwood method to produce mechanical pulp for the manufacture of cartonboard — which is coated with imported chemical pulp. The main species used is fresh-felled spruce in 1.8 to 2.3 metre lengths of 70 to 350 mm diameter, but a proportion of other softwoods is accepted. Total mill capacity is some 160,000 tonnes of cartonboard a year.

caledonian paper plc **LOG SPECIFICATION**

SPECIES	LENGTH	DIAMETER	FRESHNESS	CONTAMINATION	LOG QUALITY
SITKA SPRUCE AND NORWAY SPRUCE	2.70 m – 3.00 m (8′10″ – 9′10″)	6 cms – 40 cms (2⅜″ – 15½″)	DELIVERED WITHIN 6 WEEKS OF FELLING	NO PLASTIC NO CARBON NO METAL NO ROT OR STAIN	NO BRANCH PEGS NO FORKS NO BENDS
Noble Fir, Grand Fir and Western Hemlock are OK if mixed with Spruce. No Delivery may include more than 25% of these species.	Logs should be as close to the 3.0 m max. as possible without exceeding it.	These are overbark diameters.	Fresh felled from live green trees. No dead logs. If the cambium has started to stain the logs are too old.	No plastic labels, bottles or rubbish in the load. No fire damaged logs. Natural heartwood colour is OK. Incipient rot is not.	It must be possible to stack 1.50 m lengths close together.
Spruces have the white colour and long fibres required. Other species contain dark resin which clogs the grindstones.	Logs are cross cut to 1.50 m lengths for grinding. The shortest grindable length is 1.20 m. 1.50 + 1.20 = 2.70 m 1.50 + 1.50 = 3.00 m Longer logs produce waste.	Smaller logs break up in debarking. Bigger logs roll back down conveyors.	Dry logs will not debark easily. They overheat in the grinding process.	Contaminants are difficult to remove and either discolour or damage the paper.	Logs are handled at high speed and mis-shapen logs cause jams on conveyors.

Fig 13.7 Courtesy Caledonian paper plc

Fig 13.8 THE CALEDONIAN PAPER WOODHANDLING PROCESS

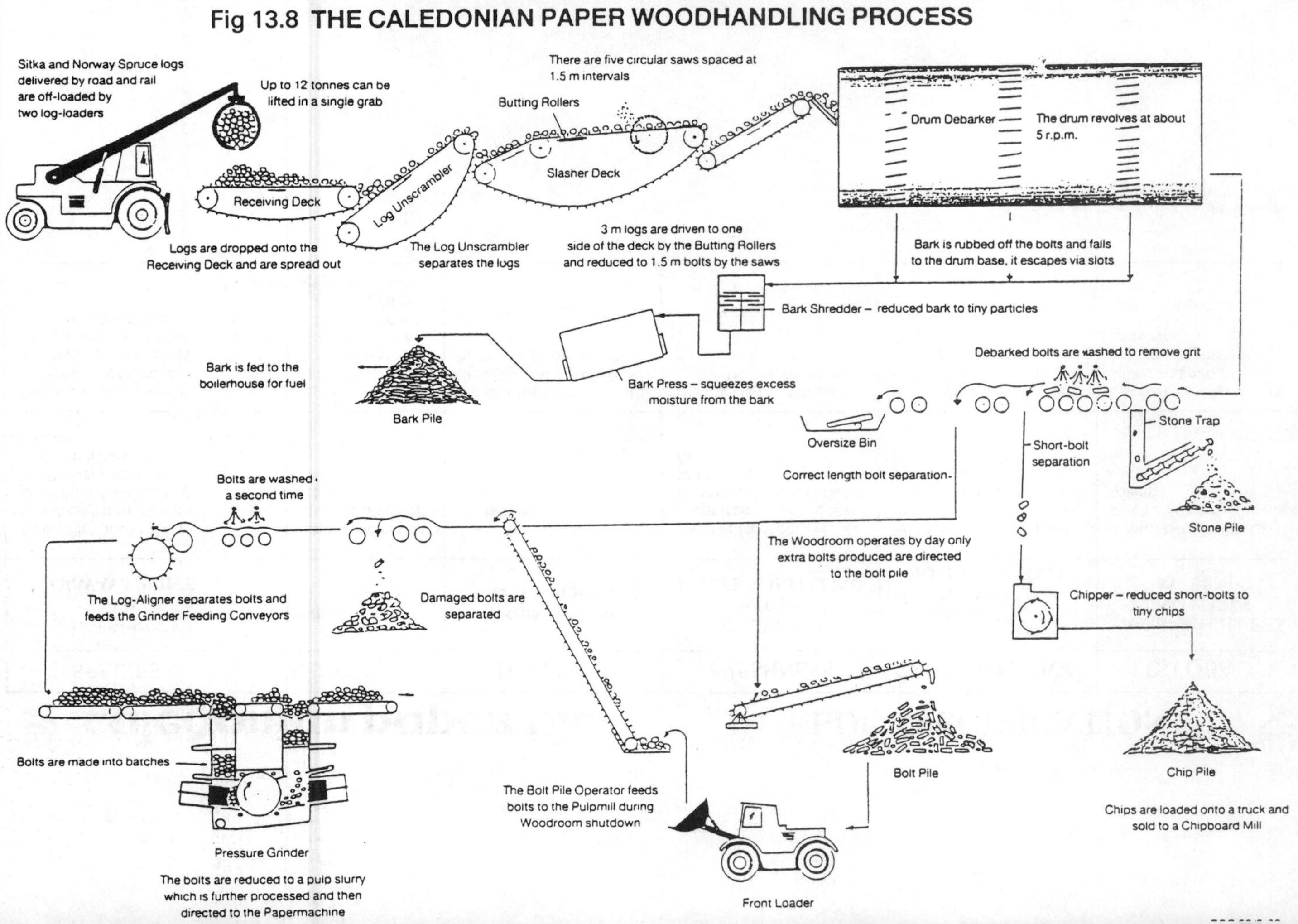

14

Wood-based Panel Products

Wood-based panel products are sheet materials which contain substantial quantities of wood in the form of veneers, chips, flakes, sawdust, fibres, strands or even small sawn blocks of uniform thickness. They include:

- plywood
- blockboard and laminboard
- particle board (also called wood chipboard)
- cement bonded particle board
- fibre building boards, which incorporate insulating board, hard-board, and semi-hardboard
- medium density fibreboard (MDF)
- oriented strand board (OSB)
- wood-wool cement slab

The above list embraces a variety of products with widely differing properties and a range of qualities within the individual product group. Nevertheless, there are a number of advantages common to all:

- Greater widths are available than are possible in sawn timber.
- The panels are manufactured in a range of standard sizes to suit the needs of the building and furniture trades.
- Individual panels are relatively homogeneous with far less variability than exists in the natural product, timber.
- Movement in changing conditions of atmospheric humidity is less than in sawn wood, and in most panel products it tends to occur to the same extent in all dimensions, i.e. it overcomes problems resulting from differential radial and tangential shrinkage.
- Species which are sometimes difficult to market can often be used for

panel products without harmful effects.

- Economic use can be made of forest, sawmill and woodworking residues in the production of several types of panel (e.g. small diameter round-wood, mis-shapen logs unsuitable for sawmilling, slabwood, off-cuts, joinery shavings and sawdust from sawmills and woodworking plants).

One of the main features of the wood and wood products market from the end of World War II has been the great increase in demand for paper and for panel products. In the UK the rate of growth of consumption of panel products has averaged 4 per cent a year since 1960; within this sector the fortunes of individual types of panel have varied widely with particle board emerging as the main product. Domestic production represents some 30 per cent of UK demand. Although it is only part of the wood raw material intake by the particle board industry, deliveries of roundwood from British woodlands have doubled in the five years to 1985.

Plywood (Veneer plywood)

The correct technical term for what is usually called plywood is "veneer plywood" because it consists of a number of wood veneers glued together, usually with the grain of adjacent veneers at right angles to each other.

Four types of plywood are commonly available:

Marine plywoods (BS1088 and 4079:1966)* are, as the name implies, used in boat building and repair. They are made normally from moderately durable (or better) timber species bonded with a phenol-formaldehyde synthetic resin.

Structural plywoods (BS5268 Pt 2 1984) are manufactured to give the finished product known strength properties.

Utility (non-structural) plywoods are intended for use in situations such as furniture and joinery where strength is not of paramount importance.

Decorative plywoods are made for special end uses such as joinery and panelling where an ornamental effect is required.

British timbers are not used continuously for plywood manufacture except for the face veneers of decorative plywood. Stout logs of many hardwoods with yew, and occasionally larch and pine, are sought by the producers of ornamental veneers both at home and on the Continent. The

* BS1088 Plywood for marine craft: Marine plywood manufactured from selected untreated tropical hardwoods. BS4079 Plywood for craft: Plywood for marine uses and treated against attack by fungi or marine borers.

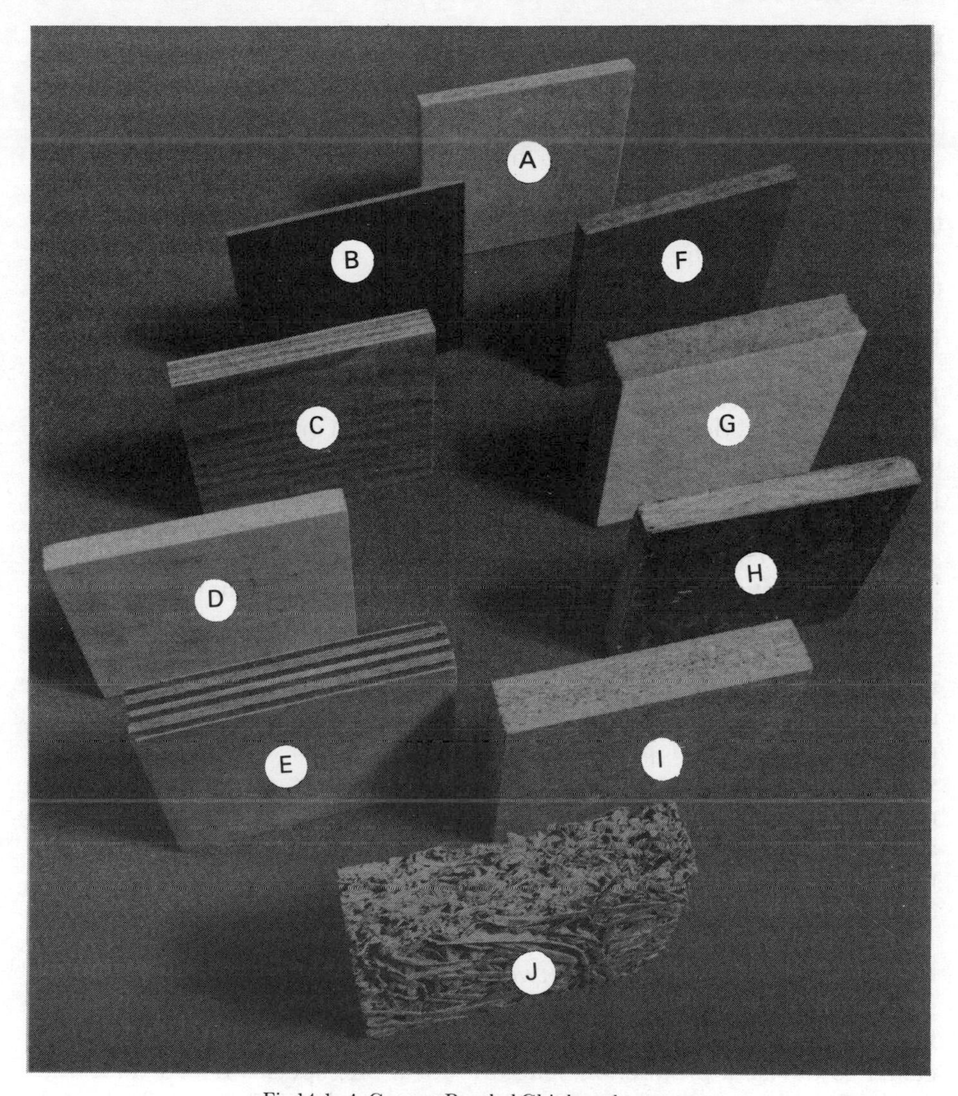

Fig 14.1 *A*, Cement Bonded Chipboard;
B, Oil-Tempered Hardboard; *C*, Scots Pine Plywood;
D, Birch Plywood; *E*, Tropical hardwood Plywood;
F, Semi-Hardboard; *G*, Insulation Board;
H, Oriented Strand Board;
I, Flooring-Grade Chipboard;
J, Wood-Wool Cement Slab.
Courtesy Building Research Station

species in regular demand include ash, cherry, oak, robinia, sweet chestnut, sycamore (especially where fiddle-back figure is known to be present), walnut and yew.

Blockboard and laminboard

While the core of a sheet of plywood is itself composed of veneers, the core of blockboard is made up of strips of wood not more than 30 mm wide, laid side by side and glued together to form a thin slab, which is then veneered on both faces. The direction of the grain of the veneers is at right angles to the long axis of the strips in the core.

In laminboard the core consists of strips of veneer on edge glued together.

Particle board (Wood chipboard)

Particle board, commonly called wood chipboard (or just chipboard) in Britain, was first developed in Germany during World War II. It is made of small chips or particles of wood, coated with a synthetic resin glue and pressed into sheets. A wide range of thicknesses, from 3 mm to 50 mm is manufactured. Because it can utilise wood residues such as planer shavings and chipped slabwood from sawmills as well as chipped small diameter forest thinnings it is less expensive to produce than plywood or blockboard which are made from higher quality woods.

Technological progress during the 1970s and 1980s has made it possible to increase the performance of particle boards and widen the number of uses, while at the same time making fuller use of wood residues. This is, in part, due to the development of technically superior adhesives, together with the development of "layered" boards where the quality of the particles improves from the core to the outer faces. There have also been numerous small but important improvements in the various stages of manufacture such as drying the particles, spreading the adhesive and the distribution of the particles in the forming press.

Where the raw material comes from a variety of species and sources, a balance may have to be made between the ease and cost of production and the performance of the board. For example since the acidity of wood varies between species, and the pH affects the setting time of the adhesives, it is easier to devise a production schedule for a one-species board than one made from a range of species. Against this, because particle board manufacturing costs per unit of board diminish as the plant size increases, it is in practice difficult to find sufficient supplies of wood of one species within an economic haulage distance from the mill.

The permeability of the particles similarly affects the production costs bearing in mind that it is desirable to retain as much of the adhesive as possible on the particle surface instead of allowing it to penetrate within where it will play no role in bonding the board. It is for this and other reasons that although manufacturers accept a variety of species, they may require some to be delivered in separately identified loads. Hence a specification for roundwood may require:

Softwoods – All British forest-grown species, but larches to be delivered under special arrangements.

Hardwoods – All British grown species, but (1) oak, sweet chestnut, and (2) poplar and willow, to be delivered under special arrangements.

The criteria by which manufacturers determine the length and diameter specifications for particle board manufacture from roundwood are similar to those used in deciding the specifications for pulpwood (see Chapter 13). In general, particle board specifications are more tolerant than those required for mechanical pulp production, since the wood can be used unbarked, or if it is debarked, total bark removal is far less critical than that required for pulp to be used in paper manufacture.

Particle board is available in long lengths, for example 12.3 by 2.255 metres (40 feet by 8 feet). Four grades are produced to meet specific end-uses; namely Standard, Flooring, Moisture Resistant and Moisture Resistant/Flooring. Moisture resistant boards will not stand prolonged wetting, nor are they suitable for use in situations exposed to the weather. The relevant British Standard is BS5669: 1979.*

The range of end-uses is very great with much of the production being acquired for furniture — including kitchen furniture and shop fittings. Flooring is a growing market. Many types of finish are available including wood and paper veneers, paint, melamine and plastic laminates.

Cement bonded particle board

Cement bonded partical board which is suitable for exterior use is not currently (1989) produced in Britain. Building blocks (which, of course, are not strictly panel products) containing sawdust and other residues have been manufactured from time to time. The early development of wood/cement materials is briefly summarised below under Wood Wool Cement Slabs.

Fibre building boards

Fibre building boards differ from plywood, blockboard and particle board in that the primary bonding of the board is derived from the inherent cohesive properties of the wood fibres when they are felted together. In the manufacturing process the fibres are most commonly produced by feeding wood chips (made from either roundwood or residues), which have been softened by pre-heating with water or steam, into a machine known as a defibrator where they are forced between grinding discs. This produces a porridge-like stock of fibres suspended in water to which substances to

* BS5669: 1979 Specification for chipboard and methods for tests for particleboard.

enhance particular properties such as moisture resistance or fire retardancy may be added. The excess water is drained away by gravity, suction and the use of pressure rollers to produce a board which is finally dried in a heated press. Pulped waste paper is sometimes also used in the feedstock.

There are four main types of fibre board (BS1142) with different uses:

Insulating Board commonly called insulation board, is a low weight building commodity with a density of not less than 240 Kg/m^3, available in thicknesses from 10 mm to 25 mm. While it has good heat and sound insulation properties, it is soft and vulnerable to accidental damage. It does not take an attractive finish and is, therefore, normally used in "concealed" situations.

Standard hardboard, produced by pressing at higher temperatures and pressures than used in the manufacture of insulation board, has a density of not less than 800 Kg/m^3. Thicknesses range from 2 mm to 6.4 mm. It is used for partition walls, bulkheads and furniture.

Tempered hardboards have densities exceeding 960 Kg/m^3. They have superior strength properties and good moisture resistance. Thicknesses range from 3.2 to 12.7 mm. They are used for structural purposes, motor vehicle panels and the external panels of caravans.

Medium density boards, known also as *semi-hardboards*, have a hardness and texture falling between insulating and standard boards. In the density range 350–560 Kg/m^3, they are used, for example, for indoor notice boards because of the ease with which they accept and retain drawing pins, and for linings. In the density range 560–800 Kg/m^3 and thicknesses from 8 to 12 mm they are used for wall linings, partitions and sheathing in timber frame housing.

Medium density fibreboard (MDF)

This relatively new product is made by adding a synthetic resin to the fibreboard feedstock after defibration. Not only is it superior to other fibre boards in terms of surface texture and smoothness, but it can be machined to very fine tolerances. It is therefore in demand for furniture production, joinery and pattern making. BS6100 defines the product and adds that the density usually exceeds 600 Kg/m^3, and that thicknesses range from 6 to 50 mm.

Oriented strand board (OSB)

Oriented strand board is also a relatively new product. It is made by coating thin strands and flakes of wood, principally Scots pine, with a phenolic resin adhesive and with wax. These are laid on a conveyor belt where they are oriented mechanically so that alternate layers are at right angles (hence the term oriented strand board). This gives the board properties comparable to plywood, although the appearance is quite dissimilar. The mat is then guided on to cauls and pressed into boards in a multi-daylight press.

Uses for oriented strand board include wall sheathing (9 mm thick); roofing (11 and 15 mm thicknesses) and tongue and groove flooring (15 and 18 mm thicknesses). For these uses the product has a British Board of Agrément certificate. Other important uses include packaging, formwork, portable buildings and DIY.

Wood wool cement slabs

It has been known for many years that the presence of organic matter in a normal cement/mineral aggregate adversely affects the strength of the final product; yet a number of useful and successful commodities, such as building boards and building blocks have been made by the controlled mixing of organic matter with cement mortars. There was, for example, a substantial production of such materials in Germany during the Second World War, using sawdust or wood splinters or wood wool as the aggregate. In those pioneering days and right into the 1970s the chemistry of the inhibiting effect of wood on the setting of cement and the formation of maximum strength were not known in any detail, but it was believed that wood sugars and the hemicelluloses were the likely sources of the problem. The palliative (used from the earliest phase of the industry) was to soak the wood particles or strands in a dilute solution of one of what were called "mineralising fluids". Calcium chloride, sodium silicate and aluminium sulphate were the preferred materials in Europe, North America and the USSR respectively. The mechanism of the setting of cement is complex but the role of the additive is to increase the rate of setting of the cement in contact with it, i.e. around the particles of wood (or other organic substances) and thus prevent the wood from affecting the setting of the cement not in intimate contact with it. Different species of wood contain different proportions of substances which cause problems with the setting of cement and so manufacturers have to be selective in the species of wood which they use.

There have been considerable advances in the understanding of the complex chemistry of wood and in the process of the setting of cement in

Fig 14.2 Wood wool cement slabs under tiled roof. *Torvale Building Products Ltd.*

the presence of organic material such as wood, but the basic principles which were established empirically still apply. The wood wool from which the slabs are made is shredded from short roundwood billets which have the bark removed prior to them being stacked for drying down to a moisture content of 20 per cent of the oven dry weight. Fig 14.2 shows Woodcemair woodwool cement slabs used for a pitched roof which was subsequently tiled.

15

Organisations Involved in the Marketing and Use of Wood

The Home Grown Timber Advisory Committee

Before listing the organisations involved in the marketing and use of wood, it is necessary to draw attention to the role of the Statutory body with special responsibilities in this field, namely the Home Grown Timber Advisory Committee.

The Forestry Commissioners in pursuance of the obligation laid upon them by Sections 37 and 38 of the Forestry Act 1967, and after consultations with the organisations appearing to them to represent the interests of owners of woodlands and of timber merchants, appoint the members of the Home Grown Timber Advisory Committee for the purpose of advising the Commissioners as to the performance of their functions under Section 1(3) and Part II of the Act and their functions of promoting the interests of forestry and the production and supply of timber.

The terms of reference of the Committee shall be to advise the Forestry Commissioners as to the performance of their functions under Section 1(3) and Part II of the Forestry Act 1967 and such other functions as the Commissioners may from time to time determine, as provided for in Section 37 (1) of the Act. These "other functions" shall include promoting the interests of forestry, the production and supply of timber and such functions as are relevant to the Commissioners' duties and responsibilities under the Countryside Acts, other Acts affecting forestry, Government policies announced from time to time, and Regulations including those emanating from the European Economic Community. More specifically, the Committee shall:

1. advise on the development of forestry both in the public and private sectors, obtaining and reviewing appropriate statistics, and considering plant health aspects;

2. advise on timber harvesting and marketing and the development of new and existing industries and markets, including appropriate research, to ensure the most effective utilisation of the country's timber resource and to contest any unreasonable discrimination against it;
3. advise on the relationship between forestry and agriculture, particularly with a view to good land use, and on environmental, recreation, nature conservation, and planning considerations associated with forestry;
4. advise on the formulation and implementation of appropriate regulations under the Forestry Act 1967 and other Acts providing for the Forestry Commission to make regulations affecting forestry, including regulations arising from membership of the EEC;
5. make an annual report on its work.

Membership of the Home Grown Timber Advisory Committee in 1987* (Appointed for a three-year period).

With the exception of seven independent members, the membership was drawn from nominations received from the following bodies: Timber Growers United Kingdom, British Timber Merchant's Association (England and Wales), Home Timber Merchant's Association of Scotland (now the United Kingdom Softwood Sawmillers' Association), British Coal, Timber Trade Federation, UK and Ireland Particleboard Association, British Paper and Board Industry Federation, and Timber Research and Development Association.

The Home Grown Timber Advisory Committee has two Sub-committees:

1. Technical Sub-committee *
   - (i) To advise the Home Grown Timber Advisory Committee on the technical aspects of:
     - (a) Harvesting and other forest operations
     - (b) Transport
     - (c) Sawmilling and other processing
     - (d) Utilisation of forest products
   - (ii) To examine, exchange and disseminate technical information within the forest industry.
2. Sub-committee on Supply and Demand *
   - (i) To compile, interpret and disseminate statistics of supply and demand for home-grown timber leading to both long-term and short-term assessments of the market.
   - (ii) To report to the Home Grown Timber Advisory Committee.

* These committees are serviced by the Forestry Commission HQ at 231 Corstorphine Road, Edinburgh, EH12 7AT.

Marketing

Produce from British Woodlands is marketed in several different ways. These include:

- sales of trees standing in the woods; the purchaser undertakes to arrange the felling and extraction.
- sales of felled trees at stump, at forest roadside or depot.
- sales of converted produce, eg sawlogs; pulpwood; poles at forest roadside or depot.

Some 57 per cent of all timber removed from the forest (removals in the mid-1980s) came from Forestry Commission woodlands, making the Commission the largest single seller of produce. The Commission sells timber standing, at roadside and delivered. Sales are by sealed tender, or at public auction, or, for a few long term contracts — mainly to pulp and board mills — by private treaty.

In the private sector, some individual owners sell round timber cut to specification at roadside or delivered. This method is common where an estate is administered for the owner by a management company operating on a regional or national basis.

The reason for this is that a company managing woods over a wide area can build up a marketing expertise, and by putting together the produce from a wide spread of estates, can offer large users of wood such as pulp and board mills, worthwhile quantities over a long period. However, most of the timber from private estates is sold standing, especially the larger trees which require specialised felling, extraction and delivery equipment, as well as a detailed knowledge of the markets.

Roundwood merchants — who purchase timber standing or felled — operate over wide areas, and their marketing expertise lies in knowing exactly where to place a particular size and quality or sawlog (or veneer butt) of an individual species. Their skills can be particularly important in marketing mature broadleaved logs to best advantage.

The sawmiller who purchases standing or felled timber may also act as a roundwood merchant, particularly for broadleaves, converting most of his purchases in his own sawmill, but selling to other consumers logs of the sizes, species and qualities in which they specialise.

Roundwood merchants and sawmillers involved in the handling of roundwood also perform the service of bringing together for bulk markets the produce from many scattered woodlands, often supplementing private woodland produce with that from Forestry Commission Woodlands.

In Britain woodland owners have seldom come together to form organisations in the way that owners have done in the Nordic countries, where major contracts (e.g. for delivery to pulp mills, or for the export of pit props) may be negotiated by a forest owners' association, which has legal powers to bind its members to fulfil contracts entered on their behalf. At home, in spite of many years of effort by small groups of enthusiasts, only a few owners' cooperatives have flourished. Nevertheless a positive attitude to co-operation in the selling of timber, without the formal establishment of an association, is now evident in some areas where the Forestry Commission, the Countryside Commission (England and Wales) and Timber Growers UK Ltd., working together at regional level have helped to co-ordinate sales of timber, particularly from farm woodlands. Furthermore, one firm of estate agents has initiated auction sales of privately owned timber enabling consumers to bid for parcels of wood from a wide area.

Organisations with a knowledge of the sale of standing trees and felled timber

Timber Growers UK Ltd.
Agriculture House
Knightsbridge
London SW1X 7NJ
and at
6 Chester Street
Edinburgh EH3 7RD

Country Landowners Association
16 Belgrave Square
London SW1X 8PQ

Scottish Landowners Federation
18 Abercrombie Place
Edinburgh EH3 7RD

Institute of Chartered Foresters
22 Walker Street
Edinburgh EH3 7HR

Publishes lists of members in consultancy including firms and individuals specialising in the measuring, valuation and marketing of timber.

Association of Professional Foresters of Great Britain
Brokerswood House
Brokerswood
Westbury
Wilts BA13 4EH

Auctioneers
(*Professional association incorporated in RICS, see below*)

Members conduct public auctions at rates agreed with the seller, depending on the amount of work involved (eg measure, describe for purposes of sale — a few firms undertake to auction parcels measured and described by the owner). Some specialise in auctioning coppice, especially sweet chestnut coppice in south east England.

Royal Institution of Chartered Surveyors
12 Great George Street
London SW1P 3AD

Processors and users of woodland produce

The main trade organisations representing the processors of British woodland produce are the two home timber merchants' associations, and the associations for the pulp and panel products industries (particle board, fibre board and oriented strand board).

Sawmillers
British Timber Merchants Association (England and Wales)
c/o Timber Research and Development Association (TRADA)
Stocking Lane, Hughenden Valley
High Wycombe
Bucks HP14 4ND

United Kingdom Softwood Sawmillers Association (formerly the Home Timber Merchants' Association of Scotland)
16 Gordon Street
Glasgow G1 3QE

Wood processors
British Paper and Board Industry Federation
3 Plough Place
Fetter Lane
London EC4 1AL

United Kingdom and Ireland Particle Board Association
Stocking Lane
Hughenden Valley

High Wycombe
Bucks HP14 4ND

United Kingdom Wood Processors' Association
Secretariat c/o Highland Forest Products
Morayhill
Inverness IV1 2JQ

Research and Development

Research into and development of the use of British grown timber is undertaken by or through Government agencies and trade associations. These include:

British Wood Preserving Association
No 6 The Office Village
Stratford
London E15 4EA

Building Research Establishment
Garston
Watford WD2 7JR

Forestry Commission
231 Corstorphine Road
Edinburgh EH12 7AT

Timber Research and Development Association (TRADA)
Stocking Lane
Hughenden Valley
High Wycombe
Bucks HP14 4ND
Publishes list of consultants in wood technology

Other bodies

There are many other organisations with an interest in the utilisation of British woodland produce. They include the following:

Association of Woodturners in Great Britain,
c/o 5 Kent Gardens,
Eastcote,
Ruislip
Middlesex HA4 8RX

British Standards Institution
Linford Wood
Milton Keynes, Bucks MK14 6LE

British Woodcarvers Association
c/o White Knight Gallery
28 Painswick Rd
Cheltenham, Glos. GL50 2HA

British Wood Turners Association
South Newton
Salisbury
Wilts SP2 OQL

Fibre Building Board Association (FIDOR)
1 Hanworth Road
Feltham
Middlesex TW13 5AT

Furniture Industry Research Association (FIRA)
Maxwell Road
Stevenage
Herts SG1 2EW

Rural Development Commission
141 Castle Street
Salisbury, Wilts SP1 3TP

Institute of Wood Science
Stocking Lane
Hughenden Valley
High Wycombe
Bucks HP14 4ND

Research Organisation for Paper and Board, Printing
and Packaging Industries (PIRA)
Randalls Road
Leatherhead
Surrey KT22 7RW

NB An annual list of key organisations, associations, scientific bodies and research institutions is published as the *Reference Issue of the Timber Trades Journal and Wood Processing*, by Benn Publications, Sovereign Way, Tonbridge, Kent TN9 1RW.

Appendix 1

Classification and Presentation of Softwood Sawlogs

The Forestry Commission has kindly given permission for the relevant sections of their publication *Classification and Presentation of Softwood Sawlogs* to be reproduced here. The full publication — which includes details of the cross-cutting charges current at the time — may be obtained direct from the Forestry Commission, Alice Holt Lodge, Wrecclesham, Farnham, Surrey, GU10 4LH.

INTRODUCTION

1. This publication replaces *Softwood Sawlogs — Presentation for Sale* which was otherwise known as Planning and Procedure Paper No 5 and which was first published in 1980. In common with the earlier publication this leaflet is based on recommendations of a joint working party of the Forestry Commission, the British Timber Merchants' Association (England and Wales) and the United Kingdom Softwood Sawmillers' Association. The text has been agreed by all three of these bodies.

CLASSIFICATION POLICY

2. The normal practice of the Forestry Commission will be to classify parcels of sawlogs offered for sale. Two categories, Green and Red, will cover all sales of logs. The description of each of these classes is given in Table 1. It will be the aim of the Commission at all times to maximise the proportion of Green category logs subject to appropriate stand conditions and market requirements.

INSPECTION

3. Potential purchasers of log parcels will have the opportunity of examining the stand or stands from which the logs will be produced before any cross-cutting is carried out. They will also be given details, including category, of any other log parcels to be taken from the same stand. In exceptional cases where logs are already cut, this will be made clear in the sale particulars.

MEASUREMENT

4. All sales will be on an underbark basis. The top diameter method of measurement (Forestry Commission Fieldbook No 1, formerly FC Booklet No 31) will be used for log lengths up to and including 8.3 m. For lengths of 8.4 m and greater, volume will be assessed by the mid-diameter method (Forestry Commission Booklet No 26) and the resulting overbark volume converted to an underbark volume using the appropriate conversion factor [see below].

5. The appropriate measurement conventions described in each of these publications will be used. Lengths are measured on the shortest side.

CROSS-CUTTING

6. It will be normal practice for lengths up to and including 8.1 m to be cross-cut in steps of 0.3 m adding an absolute minimum of 0.05 m for subsequent cross-cutting or squaring. For lengths of 8.4 m or greater, normal practice will be to cut truly random lengths. The Commission will endeavour to meet alternative length requirements to a maximum of 8.3 m, within the following options. In appropriate cases as defined below an additional charge will be levied over and above the bid price.

(a) Logs will be cut to no more than 3 preferred lengths in 0.3 m steps (or 0.1 m steps by request) plus 0.05 m cutting allowance, within the limits of log volume maximisation with a balance being produced in other lengths which fall within the description of logs offered for sale. *– no charge*

(b) Logs may be cut to *stated lengths only*, in steps of 0.1 m. An absolute minimum cutting allowance of 0.5 m will be added.
– *a charge may be levied*

SUB-STANDARD LOGS

7. Where a parcel of logs fails to meet the criteria listed for the Red Category sawlogs, for example, due to the presence of metal, the sale particulars will describe any such defficiency. These logs will be held to satisfy the criteria for the Red Category logs in all respects other than the listed deficiencies.

CONVERSION FACTORS

Underbark Volume = Overbark Volume multiplied by the conversion factor.
The agreed conversion factors are as follows:

Species	*Conversion Factor*
Scots pine, Lodgepole pine	.87
Corsican pine	.83
Sitka spruce, Norway spruce, Grand fir, Noble fir	.92
European larch	.82
Japanese larch, Hybrid larch	.85
Douglas fir	.88
Western hemlock, Red cedar, Lawson cypress	.90

Table 1 The requirements for logs classified under the Forestry Commission system

Log Category	Green	Red
Species	Any conifer – species to be stated.	
Minimum top diameter	To be stated. Normally 16 cm but not less than 14 cm. (12 cm in certain localities.)	
Length	Minimum – 1.8 m Maximum – 8.3 m	Minimum – 1.8 m Maximum – to be stated.
Cross-cut steps	0.3 m normal practice to 8.1 m maximum. (0.1 m by request, to 8.3 m maximum)	0.3 m normal practice to 8.1 m maximum. (0.1 m by request, to 8.3 m maximum) (Truly random for longer lengths.)
Straightness	Bow not to exceed 1.0 cm for every 1.0 m length and this in one plane and one direction only. Up to 5% of individual logs in any one load can be outside the specification to the extent that bow may be up to 1.5 cm for every 1.0 m length. Bow is measured as the maximum deviation at any point of a straight line joining the centres at each end of the log from the actual centre line of the log.	Capable of being cross-cut into straight lengths of at least 1.2 m without significant waste.
Knots	On any individual log 80% of knots will not exceed 5 cm in diameter. However up to 5% of the logs will be allowed out with this specification but those of an excessively coarse appearance, will be excluded.	No restriction on knot size and frequency.

Trim	For manual felling, root spurs well dressed, felling cuts as square as practicable, snedding flush to stem. With mechanised harvesting, exactly the same standards of snedding may not prove practicable, but only a modest relaxation will be acceptable. Splits, tear-outs, and double-tops are not permitted.
Scars/decay	Significant visible decay and significant scars will not be permitted.
Insect damage and staining	Visible insect damage or staining indicating incipient decay will not be present when made available for loading.
Blue stain	To minimise the infection of pine logs with blue stain, logs will be brought to the loading point within 4 weeks of felling. In view of this undertaking, the Commission will not accept blue stain as a defect or entertain claims in respect of it.
Metal	Logs suspected of containing metal will not be included.
Mean diameter at breast height	The mean dbh of *all* the standing trees which are to be removed by felling or thinning and from which the parcel of logs are to be taken, should be stated.

Appendix 2

Key British Standards of Relevance to British grown Wood

During this decade (1990's) widespread changes will be made in national standards and regulations throughout the European Economic Community, to assist in the removal of technical barriers to trade between members. The creation of harmonised European standards is being initiated by the Comité Européen de Normalisation (CEN) — on which the United Kingdom is represented by the British Standards Institution — which in addition to the EEC, includes the European Free trade area (EFTA) nations.

In pursuing its objective of harmonisation the CEN will, where possible, make maximum use of ISO (International Organisation for Standardisation) standards. When such pan-European standards are published they become mandatory in all member states, which are not then permitted to retain differing national standards or regulations for the same product or its application.

Individual CEN committees have been established to deal with sawn timber and sawlogs; structural timber; wood preservation; wood based panels; and adhesives.

Some of the current BSI standards relating to British grown wood are listed below:

* Code of practice for the structural use of timber:
– Part 2: 1984 Permissible stress design; materials and workmanship.

* Glossary of terms relating to timber and woodwork BS 565:1972 which has been replaced by BS 6100:1984 and BS 6100:1985 Glossary of building and civil engineering terms:
– Timber terms Sections 4.1 and 4.2 (1984)
– Wood based panel products Section 4.3 (1984)

– Carpentry and joinery Section 4.4 (1985)
* Glossary of terms relating to timber preservation BS 4261:1985
* Nomenclature of commercial timbers BS 881 and BS 589 (1974)
* Pallets for materials handling. Dimensions, construction and marking BS 2629: Part 1: 1967, Amendment 2721: September 1978
* Structural use of timber BS 5268.
  This includes:
  – Part 2:1989 Design, material and workmanship
  – Part 3:1985 Trussed rafter roofs
  – Part 5: 1977 Preservative treatment for constructional timbers
* Timber for and workmanship in joinery BS1186
  – Part 1: 1986 Specification for timber
  – Part 2: 1988 Specification for workmanship
* Timber grades for structural use BS4978: 1988
* Woodblocks for floors BS 1187:1959
* Wood fences and fences with some components of wood BS 1722.
This includes:
  – Part 2:1973 Woven wire fences
  – Part 3:1986 Strained wire fences
  – Part 4:1986 Cleft chestnut pale fences
  – Part 5:1986 Close boarded fences
  – Part 6:1986 Wooden palisade fences
  – Part 7:1986 Wooden post and rail fences
  – Part 11:1986 Woven and lap boarded panel fences.
* Wooden gates BS 4092:Part 2:1966
* Wood poles for overhead power and telecommunication lines: softwood poles BS 1990: Part 1: 1984. Amendment 5043:1986.

An up to date catalogue of British Standards is available at public libraries.

* For mining timber purchases British Coal use BS5750 — Quality Systems

Bibliography

AARON, J. R. (1982) *Conifer Bark: Its Properties and Uses* 2nd Edn. Forest Record 110 HMSO.

AARON, J. R. (1980) *The Production of Wood Charcoal in Great Britain* Forest Record 121 HMSO.

AARON, J. R. (1970) *The Utilisation of Bark* 2nd Edn. Res. and Dev, Paper No 32 Forestry Commission.

AARON, J. R. (1969) *Pros and Cons of Pruning in Conifers* Quart. Journ. Forestry LXIII (4).

AARON, J. R. AND OAKLEY, J. S. (1985) *The Production of Poles for Electricity Supply and Telecommunications* Forest Record 128 HMSO.

AARON, J. R. AND PRUDEN, J. J. (1970) *Experiments on Drying and Scaling Close-Piled Pine Billets at Thetford* Forest Record 72 HMSO.

AGATE, ELIZABETH (1986) *Fencing* British Trust for Conservation Volunteers.

ANON (1965) *A Comparative Study of the Properties of European and Japanese Larch* For. Prod. Res Special Rep. No 20.

ANON (1952) *An Atlas of End Grain Photomicrographs* For. Prod. Res Bull 26 HMSO.

ANON (1988) *Coniferous and Broadleaf Tree Sawlogs — Visible Defects — Classification* ISO 4473.

ANON (1989) *Forestry Facts and Figures 1987–88* Forestry Commission.

ANON *Home Grown Timbers, Corsican Pine* (1972), *Douglas Fir* (1964), *Larch* (1967), *Lodgepole Pine* (1968), *Scots Pine* (1965), *Sitka and Norway Spruce* (1967), Building Res. Est. DOE, HMSO.

ANON (1960) *Identification of Hardwoods: A Lens Key* 2nd Edn. For. Prod. Res Bull 25 HMSO.

ANON (1982) *The Use of Creosote Oil for Wood Preservation* Leaflet 8, Brit. Wood Pres. Assn.

ANON *Classification and Presentation of Softwood Sawlogs.* Field Bk 9, HMSO.

AUGUSTIN H. AND PULS, J. (1982) *Perspectives on the production of Chemicals from Wood, Chemical Processing of Wood,* Suppl. 13 to Timber Bull. for Europe XXXIV.

BANKS, W. B. (1970) *The Effect of Temperature and Storage Conditions on the Phenomenon of increased Sapwood Permeability brought about by Wet Storage* Journ. Inst. Wood Sci. 5(2).

BEGLEY C. D. AND HOWELL, R. S. (1960) *Air Seasoning of Softwoods at Stump* Forestry XXXIII (2).

BRAVERY, A. F. (1971) *The Application of Electron Microscopy in the Study of Timber Decay* Journ. Inst Wood Sci. 5(6).

BRAZIER, J. D. (1985) *Growing Hardwoods: The Quality Viewpoint* Quart Journ For. 79(4).

BRAZIER, J. D. AND MOORE, G. L. (1985) *British Grown Timber of* Nothofagus obliqua *and* Nothofagus procera Quart Journ For. 79(2).

BRAZIER, J. D., HANDS, R. G. AND SEAL, D. T. (1985) *Structural Wood Yields from Sitkas Spruce: The Effects of Planting Spacing* For. and Brit Timb. Sept.

BRITISH STANDARD 1186 (2 parts) *The quality of timber and workmanship in joinery.*

—— 1722 (13 parts) *Fences.*

—— 1990:1984 *Wood poles for overhead power and telecommunication lines.*

—— 2482:1981 *Timber scaffold boards.*

—— 2548:1986 *Wood wool for general packaging purposes.*

—— 2629 (3 parts) *Pallets for materials handling for through transit.*

—— 3819:1964 (withdrawn) *Grading rules for sawn British softwood.*

—— 4047:1966 (withdrawn) *Grading rules for sawn British hardwood.*

—— 4169:1970 *Glued laminated timber structural members.*

—— 4978:1988 *Softwood grades for structural use.*

—— 5268:1984 (part 2) *Structural use of timber: Code of practice for permissible stress, design and workmanship.*

—— 5291:1984 *Manufacture of finger joints in structural softwoods.*

—— 5502:1986 (3 parts) *Code of practice for designs of buildings and structures in agriculture.*

—— 5696 (3 parts) *Play equipment intended for permanent installation outdoors.*

—— 5750 (6 parts) *Quality systems.*

—— 6100:1984 *Building and civil engineering terms: Pt 4 Forest products.*

BROWN, Wm. (1986) *Conversion and Seasoning of Wood*, Stobart Davies.

CHITTENDEN, A. E., HAWKES, A. J. AND HILARY R. HAMILTON (1974) *Wood Cement Systems* FO/WCWP p/75 Doc. 99 FAO Rome.

CROWTHER, R. E. AND PATCH, D. (1980) *How Much Wood for the Stove* Res. Inf. Note For. Comm.

DINWOODIE, J. M. (1978) *The Properties and Performance of Particleboard Adhesives* Journ. Inst. Wood Sci. 8(2).

DADSWELL, H. E., WARDROP, A. B. AND WATSON, A. J. (1958) *The Morphology, Chemistry and Characteristics of Reaction Wood* Fundamentals of Paper-Making Fibres pp. 187–229, Technical sect. Brit Paper and Board Makers Assn.

DUNLEVY, J. A. AND MCQUIRE, A. J. (1970) *The Effect of Water Storage on the Cell Structure of Sitka Spruce (*Picea sitchensis*) with Reference to its permeability and preservation* Journ. Inst. Wood Sci. 5(2).

EDLIN, H. L. (1969) *What Wood is That? A Manual of Wood Identification* Stobart Davies.

FORSS, K. (1982) *Chemical Processing of Wood in Finland* Chemical Processing of Wood, Suppl. 13 to Vol XXXIV of the Timb. Bull. for Europe.

GARETH JONES, E. R. AND IRVINE, A. (1971) *The Role of Fungi in the Deterioration of Wood in the Sea* Journ. Inst. Wood Sci. 5(5).

GRAYSON, A. J. (1989) *Carbon Dioxide, Global Warming and Forestry* Res. Inf. Note 146 For. Comm.

GREAVES, H. AND LEVY, J. F. (1965) *Comparative Degradation of the Sapwood of Scots Pine, Beech and Birch by Lenzites trabea, Polystictus versicolor, Chaetomium globosum and Bacillus polymixa* Journ. Inst. Wood Sci. No 15.

HAMILTON, G. J. AND CHRISTIE, J. M. (1971) *Forest Management Tables (Metric)* For. Comm. Booklet 34 HMSO.

HASELTINE, B. A. (1975) *The Comparison of Energy Requirements for Building Materials and Structures* Struct. Eng. Sept (Discussion published Struct. Eng. June 1976).

HILLIS, W. E. (1975) *The Role of Wood Characteristics on High Temperature Drying* Journ. Inst. Wood Sci. 7(2).

HUM M, GLASER, A. E. AND EDWARDS, R. (1980) *Wood Boring Weevils of Economic Importance in Britain* Journ. Inst. Wood Sci. 8(5).

JACKMAN, P. E. (1981) *The Fire Behaviour of Timber and Wood Based Products* Journ. Inst. Wood Sci. 9(1).

JANE, F. W. (1970) *The Structure of Wood* 2nd Edn. A & C Black.

JEFFERS, J. N. R. (1966) *Relationship between Compressive Strength, Moisture Content, Rate of Growth and Maximum Bow in Home-Grown Pit-Props* Forestry 39(1).

LAVERS, G. M. (1968) *The Strength Properties of Timbers* Forest Prod. Res. Bull. No 50 2nd Edn. HMSO.

LINCOLN, W. A. (1986) *World Woods in Colour* (Includes a table of uses) Stobart Davies.

LYNAM, F. C. (1959) *Factors Influencing the Properties of Chipboard* Journ. Inst. Wood Sci. No 4.

METCALFE, C. R. AND CHALK, L. (1983) *Anatomy of the Dicotyledons* 2nd Edn. Clarendon Press.

OLIVER, A. C. (1962) *An Account of the Biology of Limnoria* Journ. Inst. Wood Sci. No 9.

PURSLOW, D. F. AND REDDING, L. W. (1978) *Comparative Tests on the Resistance to Impregnation with Creosote and Copper/Chrome/Arsenic Water Borne* Journ. Inst. Wood Sci. 8(1).

RICHARDS, E. G. (1987) *Forestry and Forest Industries: Past, Present and Future* Martinus Nijhoff for the United Nations.

RICHARDS, E. G. (1987) Special Issue Timb. Bull. for Europe 39(3). United Nations Geneva.

SMITH, D. N. AND WILLIAMS (1969) *Wood preservation by the Boron Diffusion Process: The Effect of Moisture on Diffusion Time* Timberlab Papers No 5 Princes Risborough Laboratory.

SMITHIES, J. N. AND MAUN, K. W. (1984) *LOCAS — A control System for Sawmills* B R E Inf. Paper 3/84.

SUNLEY, J. G. (1965) *The Compressive Strength of Home-Grown Pit-Props* Forest Prod. Res. Sp. Rep. No 21 HMSO.

VINDEN, P. (1984) *The Effect of Raw Material Variables on Preservative Treatment of Wood by Diffusion Processes* Journ. Inst. Wood Sci. 10(1).

WARDROP, A. B. (1962) *Fundamental Studies in Wood and fibre Structure relating to Pulping Processes* APPITA 16(3).

WARDROP, A. B. (1954) *The Fine Structure of a Conifer Tracheid* Holzforsch. 8(1).

WHITEHEAD, D. (1978) *The Estimation of Foliage Area from Sapwood Basal Area in Scots Pine* Forestry 51(2).

WILSON, I. M. (1954) Ceriosporopsis halima *Linder and* Ceriosporopsis cambrensis *sp. nov: Two Marine Pyrenomycetes on Wood* Trans Brit Myc Soc 37(3).

WILSON, K. AND WHITE, D. J. B. (1987) *The Anatomy of Wood: Its Diversity and Variability* Stobart Davies.

WILSON, S. E. (1944–45) *Mining Timber*. Reprinted from *Wood* by The Lumberbook Company, Clevedon, Somerset.

Index

Abies, 149
Acer, 163
Acland Report, 15
Activated charcoal, 131
Aesculus, 159
Air-drying, 61
Alder, 156
Alnus, 156
Ambrosia beetle, 113
Ancient and semi-natural woodland, 18
Anobium, 114, 115
Apple, 165
Armillaria, 107
Ash, 157
Association of Professional Foresters, 205
Association of Wood Turners in Great Britain, 207
Aspergillosis, 174
Aspirated pits, 182
Auctioneers, 206
Austrian pine, 148

Bacteria, 112
Bacterial digestion of pit membranes, 113
Band saw, 53, 54, 57
Bark, 171
Beech, 157
Beetle damage, 114, 116, 117
Bethell process, 121
Betula, 158
Birch, 158
Bleaching of pulp, 188
Blockboard, 196
Blue stain, 108, 109
Boiling under vacuum, 121
Bordered pits, 181
Boron diffusion, 119
Boucherie process, 34, 35
Boulton process, 121
Branches, 168
British Leather Manufacturers' Research Association, 171
British Standards Institution, 207
British Timber Merchants Association, 206
British Paper and Board Industry Federation, 206
British Wood Carvers Association, 208
British Wood Preserving Association, 207
British Wood Turners Association, 208
Broadleaf woodland, 18
Brown rot, 110
Building Regulations, 96
Building Research Establishment, 207
Butt rot, 24, 69, 74

Cable cranes, 23
Calorific values, 130
Cant sawing, 55, 56
Carbonisation, 127, 131, 132, 133
Carpinus, 160
Castanea, 158
Case-hardening, 63
CCA, 119
Cedars, 148, 149
Cellar Fungus, 69, 110, 111
Cell wall structure, 180, 181
Cellulose, 177
Ceratocystis, 25
Chamaecyparis, 148
Charcoal, 131, 134
Chard, R., 44
Checks, 63, 77
Chemical pulping, 188
Cherry, 158
Chestnut (Horse), 159
Chestnut (Sweet), 87, 158
Chipboard, 196
Chipping, 169
Chloronaphthanates, 120
Chocks, 39, 103
Christmas trees, 149, 169
Circular saws, 51, 57
Cleft chestnut fences, 87
Cleft timber, 48, 87
Coigue, 164
Collapse, 63
Coniophera, 69, 110, 111
Compression wood, 74, 75
Compressive strength, 40, 155, 166
Cooperage, 184
Copper chrome arsenate, 119
Copper naphthenate, 120
Corsican pine, 32, 144

Country Landowners Association, 205
Cover boards, 103
Creosote, 118
Cross-cutting, 24
Cross-field pits, 182
Crown trees, 103
Cutting system of hardwood grading, 78

Dead knots, 72, 73
Death watch beetle, 115, 116
Decay, 68, 106
Defects, 68
Dehumidifyers, 65
Dendroctinus, 25
Diffuse porous hardwoods, 184
Diffusion treatment with preservatives, 36, 126
Dihydroquercetin, 171
Disc-refiner pulping, 187
Disodium octoborate, 119, 126
Double vacuum process, 125
Douglas fir, 32, 146
Dry Rot, 110, 111
Dunkeld larch, 32, 145
Dunnage, 104
Durability, 112

Earlywood, 180
Economic Commission for Europe, 12, 45, 80
Electric fences, 91
Electricity supply poles, 36
Elm, 159
Empty cell process, 122
Energy plantations, 131
Epithelial cells, 181
Equilibrium moisture content, 67
Ernobius, 117
Eucalyptus, 131, 163
European cherry, 158
European Forestry Commission of the United Nations Food and Agriculture Organisation, 12, 80
European larch, 145
European plane, 163
European silver fir, 149
European spruce, 32, 140, 142
Europhryum, 117

Fagus, 157
Farmers' lung, 171, 174
Fibre Building Board Association, 208
Fibre saturation point, 59
Fibre tracheids, 184
Fibreboard, 197
Fibres, 184
Finger jointing, 97, 99
Fire resistance, 137, 138, 139
Fissures, 72, 77
Flag poles, 37
Flame retardant treatment, 137, 138
Flat sawn, 55, 57
Floristry, 169
Fomes (Heterobasidion), 69
Food and Agriculture Organisation of the United Nations, 12, 45
Forest longhorn beetles, 113, 114
Forest policy, 15, 16
Forestry Commission, 207
Forwarders, 21, 22
Frame saws, 51, 53, 57
Fraxinus, 157
Fuelwood, 128
Full cell process, 121
Full tree system, 20, 21, 22
Fungi, 106
Furfural (furfuraldehyde), 178
Furniture, 103
Furniture beetle, 114, 115
Furniture Industry Research Association, 208

Gassification, 134
Glucose, 178
Glued laminated structures, 97, 99
Grading rules, 77
Grand fir, 149
Gribble, 117
Growth rate, 76, 90
Groundwood pulp, 184, 187

Half round post and rail fence, 82
Hardboard, 198
Harvesting, 19
Hazel, 88
Heart rot, 106
Heartwood, 28
Hemicellulose, 178
Heterobasidion, 24, 69, 167
Hollocellulose, 177
Holly, 164
Home Grown Timber Advisory Committee, 202
Home Timber Merchants Association of Scotland (UK Softwood Sawmillers Association), 206
Hornbeam, 160
Horticulture, 172

Hot and cold treatment with creosote, 123
House longhorn beetle, 115, 117
Hurdles, 88
Hybrid (Dunkeld) larch, 145
Hydrolysis of wood, 178
Hylotrupes, 115, 117

Ilex, 164
Incising, 120
Infra-red drying, 65
ISO sawlog classification, 45, 46, 47
ISO stress grading and finger jointing, 80
Initial spacing of trees, 17
Insects, 113
Institute of Chartered Foresters, 205
Institute of Wood Science, 208
Insulating board, 198
Insulation board, 198
Imports, 93

Japanese larch, 145
Joinery, 96, 98
Jointed cutters, 73
Juglans, 164
Juvenile wood, 76, 77

Kiln drying, 63, 67
Knot area ratio, 78
Knot holes, 72, 73
Knots, 70
Kraft pulping process, 186

Laburnum, 164
Laminated structures, 97, 99
Landscaping, 170, 172
Larch, 32, 90, 145
Larix, 145
Latewood, 180
Lawson Cypress, 148
'League tables' for strength properties, 155, 166
Lids, 39, 103
Lignin, 178
Lime, 160
Line thinning, 18
Live knots, 72, 73
Lodgepole pine, 145
London plane, 163
Loose knots, 73, 144
Lop and top, 22
Log grading, 43
Longhorn beetles, 114, 115
Lowry process, 122
Lumen, 181
Lyctus, 115, 116

Malus, 165
Marine borers, 117
Marine fungi, 118
Maritime pine, 143
Marketing, 204
Mechanical pulping, 184, 187
Mechanical stress grading, 79
Medium density fibreboard, 198
Middle lamella, 180
Microfibrils, 180
Mining timber, 38, 39, 103
Monoterpenes, 172
Movement in wood, 66
Mould fungi, 108
Mulching, 173

Narcedes, 117
Natural durability, 112
Naval stores, 179
Noble fir, 149
Non-thin regimes, 18
Norway spruce, 142
Nothofagus, 164

Oak, 160
Oak tan bark, 171
Octoborate, 119
Open tank process, 123
Organisation for European Cooperation and Development, 12
Oriented strand board, 199

Packaging, 100
Pallets, 101, 102
Parenchyma, 184
Particle board, 196
Performance of hardwoods, 166
Performance of softwoods, 154–5
Pear, 165
Peniophora, 24
Pentarthrum, 117
Picea, 140, 141, 142
Pigeon-hole stacking, 61, 62
Pillar wood, 39, 103
Pin hole borers, 113
Pinene, 179
Pine, 32, 143, 144, 145
Pinus, 143, 144, 145
Pith, 56
Pit props, 37, 38, 39, 40, 41
Pits in cell walls, 181
Pit sawing, 49
Pitwood, 37, 38, 40, 41
Plain sawn, 55, 57

Plane, 163
Playgrade bark, 173, 174
Platanus, 163
Plywood, 194
Polarised light, 178
Poles, 31
Pole barns, 37
Polish manufacture, 171
Poplar, 162
Populus, 162
Port Orford cedar, 148
Portable sawmills, 50, 52
Post and rail fences, 81, 82, 86
Potting mixtures, 171
Poultry litter, 174, 175
Powder post beetles, 115, 116
Power transmission poles, 31,
Profile chipper, 53, 54, 56, 57
Pruning, 45
Prunus, 158
Pseudotsuga, 146
Ptilinus, 116
Pyrus, 165

Quarter sawn, 55, 57
Quercus, 160

Radiata pine, 143
Radio frequency drying, 65
Rate of (radial) growth, 76, 90
Rauli, 164
Reaction wood, 73
Red oak, 162
Redwood, 143
Resin canals, 182
Research Organisation for the Paper and Board, Printing and Packaging Industries, 208
Resin, 179
Rift sawn, 55, 57
Ring-porous, 182
Ring width, 76, 90
Robinia, 164
Roots, 167
Rosin, 179
Royal Institution of Chartered Surveyors, 206
Roundwood merchants, 204
Rueping process, 122
Rural crafts, 104
Rural Development Commission, 208
Rustic poles, 42

Salix, 164
Sap displacement methods of preservation, 33, 34, 35
Sap peeling of tan bark, 58
Sap stain, 108
Sapwood, 28, 33
Sawing parallel to the bark or pith, 57
Sawn mining timber, 103, 166
Scandinavian grading, 78
Scaffold boards, 100
Scots pine, 32, 143
Scottish Landowners' Federation, 205
Seed orchards, 18
Semi-chemical pulping, 188
Semi-ring porous, 184
Sequoia and *Sequoiadendron*, 150
Serpula, 111
Shake, 45, 46, 159
Shingles (roof), 150
Shipworm, 117
Shortwood system, 21
Shrinkage, 66
Silver fir, 149
Sitka spruce, 33, 140, 141
Sleepers, 102
Sloping grain, 73
Smelting, 131
Soft rot, 110
Softwood Sawmillers Association, 206
Sodium pentachlorphenate, 120
Solar drying, 65, 66
Sound knots, 73
Southern beech, 164
Spiral grain, 73
Splits (Defect), 63, 77
Spacing at time of planting, 17
Springwood, 180
Spruce, 32, 140
Stains, 75
Staining fungi, 75
Steeping, 126
Stereum, 106, 107
Stickers, 61
Strained wire fence, 83
Strategic reserve of timber, 15
Stump treatment, 24
Sub-committee on Supply and Demand of the HGTAC, 203
Summer felling, 58
Summerwood, 180
Surface moulds, 108
Sycamore, 163

Tan bark, 171, 172

Tannins, 172, 173
Taxus, 150
Telegraph poles, 31, 36
Tension wood, 75
Teredo, 117
Thermo-mechanical pulping, 187
Thinning, 18
Through-and-through sawing, 55
Thuja, 149
Tilia, 160
Timber frame buildings, 94
Timber Growers UK Ltd, 205
Timber Research and Development
Association, 207
Tops, 168
Torus, 181
Tracheids, 179
Traditional house building, 95
Tributyl tin oxide, 120
Tree length system, 22
Tsuga, 149
Turkey oak, 162
Turpentine, 179
Twist, 60
Tyloses, 184

UK and Ireland Particle Board
Association, 206
UK Softwood Sawmillers Association,
206
UK Wood Processors Association, 207
Ulmus, 159, 160
United Nations Economic Commission
for Europe, 12
Uronic acids, 178

Veneers, 194
Vertical (Axial) resin canals, 182
Vessels, 182
Volatile oils, 172

Walking sticks, 42
Walnut, 164
Wane, 75
Water, structural role in wood, 179
Waxes, 171
Weevils, 117
Wellingtonia, 150
Western hemlock, 149
Western red cedar, 149
Wet rot, 110, 111
Weymouth pine, 143
Wharf borer, 117
Whisky barrels, 184
White rot, 110
Whitewood, 149
Wild cherry, 158
Wild grain, 68, 159
Willow, 164
Wilson system of log grading, 44
Windthrow, 18
Wood wasps, 113, 114
Wood wool, 40
Wood wool slabs, 199, 201
Wood worm, 114, 115
Wreath making, 169
Wych elm, 160

Xestobium, 115
X-ray diffraction, 178

Yellow pine, 143
Yew, 150

Zinc naphthanate, 120